AF352827

# Carbazole Derivatives: Latest Advances and Prospects

# Carbazole Derivatives: Latest Advances and Prospects

Editor

**Anna Caruso**

MDPI • Basel • Beijing • Wuhan • Barcelona • Belgrade • Manchester • Tokyo • Cluj • Tianjin

*Editor*
Anna Caruso
University of Calabria
Italy

*Editorial Office*
MDPI
St. Alban-Anlage 66
4052 Basel, Switzerland

This is a reprint of articles from the Special Issue published online in the open access journal *Applied Sciences* (ISSN 2076-3417) (available at: https://www.mdpi.com/journal/applsci/special_issues/carbazole_derivatives).

For citation purposes, cite each article independently as indicated on the article page online and as indicated below:

LastName, A.A.; LastName, B.B.; LastName, C.C. Article Title. *Journal Name* **Year**, *Volume Number*, Page Range.

**ISBN 978-3-0365-7288-8 (Hbk)**
**ISBN 978-3-0365-7289-5 (PDF)**

# Contents

*Editorial*

# Special Issue "Carbazole Derivatives: Latest Advances and Prospects"

**Anna Caruso** [1,2]

1  Department of Pharmacy, Health and Nutritional Sciences, University of Calabria, 87036 Arcavacata di Rende, Italy; anna.caruso@unical.it or anna.caruso@unibas.it
2  Department of Science, University of Basilicata, 85100 Potenza, Italy

## 1. Introduction

The academic community has extensively explored, over the years, heterocyclic compounds of the carbazolic motif. These extremely versatile molecules have found applications both in science materials and the pharmaceutical industry. In particular, it was shown that many carbazole derivatives possess a diversity of biological activities and could be used as anticancer, antimicrobial, anti-inflammatory, antioxidant, antiepileptic, antihistamine, antidiarrheal, analgesic, antidiabetic, and neuroprotective agents. Overall, the data reported in this Special Issue could constitute an important resource for the development of novel, efficient, and safe drugs for the treatment of various diseases.

## 2. The Present Issue

A study in this Special Issue reported the $\beta$1-blocking activity of 5,8-dimethyl-9*H*-carbazol-3-yl ethyl carbonate and its derivatives. In particular, two of these derivatives, 1-methyl-1*H*-indol-5-yl-but-2-ynoate and indol-5-yl-but-2-ynoate, emerged as potential counteracting agents against ISO-dependent in vitro cardiac hypertrophy. The data are promising as they were obtained using lower concentrations than those of the traditional $\beta$-AR antagonist propranolol. Following molecular docking studies, the authors tested these molecules by bioassays in H9c2 cardiomyocytes exposed to isoproterenol (ISO) to confirm their potential as $\beta$1-blocking agents and their activity at low doses, along with their limited side effects [1].

Some studies suggest that many carbazole derivatives could be promising anticancer agents. According to some studies, their effects could also be due to the involvement of the JAK/STAT pathway. According to literature data, some carbazoles could act by downregulating STAT proteins, mostly STAT-3, also affecting interleukins and *i*-NOS production. Several researchers have evaluated the STAT inhibitory activity of different carbazoles, such as carbamazepine, 2-hydroxycarbazole, mahanine, 7-hydroxy-1-methyl-9*H*-carbazol-2-yl 5-(dimethylamino)-naphthalene-1-sulfonate, EC-70124, Lestaurtinib, and some series of *N*-alkylcarbazoles, 1,4-dimethyl-carbazoles, 9*H*-carbazole-1-carboxamides, and carbazol carbonitriles, reporting promising results [2]. Other important work has reported the anticancer activity of a series of nitrocarbazoles. One of the latter compounds, namely 2-nitro-1,4-di-*p*-tolyl-9*H*-carbazole, exhibited good anticancer activity against two breast cancer cell lines, i.e., ER(+) MCF-7 and triple-negative MDA-MB-231, with $IC_{50}$ values of $7 \pm 1.0$ and $11.6 \pm 0.8$ µM, respectively. Furthermore, this compound did not interfere with the growth of the normal cell line MCF-10A (human mammary epithelial cell line). In vitro immunofluorescence studies and docking simulations demonstrated the ability of the 2-nitrocarbazole derivative to interfere with tubulin organization, a feature that results in triggering MCF-7 cell death by apoptosis [3].

Several research groups have extensively studied the involvement of carbazole derivatives, of synthetic or natural origin, in diabetes pathways. From the analysis of these

**Citation:** Caruso, A. Special Issue "Carbazole Derivatives: Latest Advances and Prospects". *Appl. Sci.* **2023**, *13*, 4263. https://doi.org/10.3390/app13074263

Received: 16 March 2023
Accepted: 20 March 2023
Published: 28 March 2023

studies it was shown that *N*-substitution, on the carbazole scaffold, by triazinic-, as in 1-((5,6-di(furan-2-yl)-1,2,4-triazin-3-yl)thio)-3-(3,6-dibromo-9*H*-carbazol-9-yl)propan-2-ol, or triazolic moieties, as in 2-(4-((9*H*-carbazol-9-yl)methyl)-1*H*-1,2,3-triazol-1-yl)-1-(3-bromo-4-hydroxyphenyl)ethanone, favors α-glucosidase inhibitory activity; the presence of cyclic sulfonamidic, as in 2-(3-(9*H*-carbazol-9-yl)-2-hydroxypropyl)isothiazoline-1,1-dioxide, or cyclic urea groups, as in 1-(3-(3,6-difluoro-9*H*-carbazol-9-yl)-2-hydroxypropyl)imidazolidin-2-one, modulates cryptochrome activity; the presence of ethylphenoxy groups, as in Chiglitazar, improves insulin sensitivity; hydrogenation of a ring of the carbazole core promotes the hypoglycemic effect via the AMPK pathway, as in 6-(benzyloxy)-9-(4-chlorobenzoyl)-2,3,4,9-tetrahydro-1*H*-carbazole-3-carboxylic acid, and the tetracyclic system, as in mahanine derivatives, is crucial in glucose uptake and translocation of glucose transporter GLUT4 in skeletal muscle and adipocyte cells. Furthermore, compounds with a high conjugation effect, such as bisgerayafoline D, show predominant antioxidant activity. The data reported could provide an important guide reference for the development of alternative and effective antidiabetic agents [4].

Lastly, considering the pandemic period that has affected everyone's life since 2020, some authors reported the latest carbazole therapeutic strategies for Severe Acute Respiratory Syndrome Coronavirus-2 (SARS-CoV-2) infection. In particular, they reported the data of 14 drugs containing a carbazole structure, including Carprofen, Carvedilol, and five carbazole alkaloids obtained from *Murraya koenigii*, which were studied as SARS-CoV-2 M-pro inhibitors. Edotecarin, 7-hydroxystaurosporine, CIMSSNa, and 6-formylindolo(3,2-b)carbazole were reported as viral entry inhibitors targeting human ACE2, while 2-((2-(1-benzylpiperidin-4-yl)ethyl)amino)-*N*-(9*H*-carbazol-9-yl)acetamide was reported as an NPC1 inhibitor; 6-cyano-5-methoxy-12-methylindolo [2, 3A] carbazole was reported as antiviral against PLpro; and Ramatroban was reported as an immunotherapy treatment. The studies and observations which reported about carbazoles for the treatment of COVID-19 infection can signify potentially useful clinical applications. Moreover, some of these molecules can be used for designing new antivirals [5].

**Funding:** This research received no external funding.

**Conflicts of Interest:** The author declares no conflict of interest.

## References

1. Grande, F.; De Bartolo, A.; Occhiuzzi, M.; Caruso, A.; Rocca, C.; Pasqua, T.; Carocci, A.; Rago, V.; Angelone, T.; Sinicropi, M. Carbazole and Simplified Derivatives: Novel Tools toward β-Adrenergic Receptors Targeting. *Appl. Sci.* **2021**, *11*, 5486. [CrossRef]
2. Caruso, A.; Barbarossa, A.; Carocci, A.; Salzano, G.; Sinicropi, M.; Saturnino, C. Carbazole Derivatives as STAT Inhibitors: An Overview. *Appl. Sci.* **2021**, *11*, 6192. [CrossRef]
3. Sinicropi, M.; Tavani, C.; Rosano, C.; Ceramella, J.; Iacopetta, D.; Barbarossa, A.; Bianchi, L.; Benzi, A.; Maccagno, M.; Ponassi, M.; et al. A Nitrocarbazole as a New Microtubule-Targeting Agent in Breast Cancer Treatment. *Appl. Sci.* **2021**, *11*, 9139. [CrossRef]
4. Grande, F.; Ioele, G.; Caruso, A.; Occhiuzzi, M.; El-Kashef, H.; Saturnino, C.; Sinicropi, M. Carbazoles: Role and Functions in Fighting Diabetes. *Appl. Sci.* **2023**, *13*, 349. [CrossRef]
5. Bonomo, M.; Caruso, A.; El-Kashef, H.; Salzano, G.; Sinicropi, M.; Saturnino, C. An Update of Carbazole Treatment Strategies for COVID-19 Infection. *Appl. Sci.* **2023**, *13*, 1522. [CrossRef]

*applied*
*sciences*

*Article*

# Carbazole and Simplified Derivatives: Novel Tools toward $\beta$-Adrenergic Receptors Targeting

Fedora Grande [1,†], Anna De Bartolo [2,†], Maria Antonietta Occhiuzzi [1], Anna Caruso [1,*], Carmine Rocca [2], Teresa Pasqua [3], Alessia Carocci [4], Vittoria Rago [1], Tommaso Angelone [2] and Maria Stefania Sinicropi [1]

[1] Department of Pharmacy, Health and Nutritional Sciences, University of Calabria, 87036 Arcavacata di Rende, Italy; fedora.grande@unical.it (F.G.); mariaantonietta.occhiuzzi@unical.it (M.A.O.); vittoria.rago@unical.it (V.R.); s.sinicropi@unical.it (M.S.S.)
[2] Laboratory of Cellular and Molecular Cardiovascular Pathophysiology, Department of Biology, E and E.S. (DiBEST), University of Calabria, 87036 Arcavacata di Rende, Italy; anna.debartolo@unical.it (A.D.B.); carmine.rocca@unical.it (C.R.); tommaso.angelone@unical.it (T.A.)
[3] Department of Health Science, University of Catanzaro Magna Graecia, 88100 Catanzaro, Italy; teresa.pasqua@unicz.it
[4] Department of Pharmacy-Pharmaceutical Sciences, University of Bari Aldo Moro, 70126 Bari, Italy; alessia.carocci@uniba.it
[*] Correspondence: anna.caruso@unical.it; Tel.: +39-0984-493019
[†] These authors equally contributed to this work.

**Abstract:** $\beta$-Adrenergic receptors ($\beta$-ARs) are G protein-coupled receptors involved in important physiological and pathological processes related to blood pressure and cardiac activity. The inhibition of cardiac $\beta$1-ARs could be beneficial in myocardial hypertrophy, ischemia and failure. Several carbazole-based compounds have been described as promising $\beta$-blockers. Herein, we investigate the capability of a carbazole derivative and three simplified indole analogs to interact with the active binding site of $\beta$1-AR by molecular docking studies. In the light of the obtained results, our compounds were tested by biological assays in H9c2 cardiomyocytes exposed to isoproterenol (ISO) to confirm their potential as $\beta$1-blockers agents, and two of them (**8** and **10**) showed interesting and promising properties. In particular, these compounds were effective against ISO-dependent in vitro cardiac hypertrophy, even at concentrations lower than the known $\beta$-AR antagonist propranolol. Overall, the data suggest that the indole derivatives 8 and 10 could act as potent $\beta$1-blockers and, active at low doses, could elicit limited side effects.

**Keywords:** $\beta$-blockers; carbazole; indole; molecular docking; H9c2 cells; cardiac hypertrophy

**Citation:** Grande, F.; De Bartolo, A.; Occhiuzzi, M.A.; Caruso, A.; Rocca, C.; Pasqua, T.; Carocci, A.; Rago, V.; Angelone, T.; Sinicropi, M.S. Carbazole and Simplified Derivatives: Novel Tools toward $\beta$-Adrenergic Receptors Targeting. *Appl. Sci.* **2021**, *11*, 5486. https://doi.org/10.3390/app11125486

Academic Editor: Ana M. L. Seca

Received: 28 April 2021
Accepted: 11 June 2021
Published: 13 June 2021

**Publisher's Note:** MDPI stays neutral with regard to jurisdictional claims in published maps and institutional affiliations.

## 1. Introduction

Over the last decades, $\beta$-blockers (i.e., antagonists of $\beta$-adrenergic receptors, $\beta$-ARs) have emerged as crucial therapeutic agents in the first-line treatment of acute and chronic diseases, particularly in the field of cardiovascular pathologies.

$\beta$-ARs are G protein-coupled receptors used by endogenous catecholamines, especially noradrenaline and adrenaline [1], to modulate important physiological processes, such as blood pressure and cardiac activity [2]. Starting from their discovery, which gained Sir Henry H. Dale the Nobel Prize in 1936 [3], $\beta$-ARs have become crucial targets in the therapy of cardiovascular diseases [4]. Indeed, it is widely accepted that a deep relationship between $\beta$-ARs and the cardiovascular role of the sympathetic nervous system (SNS) exists. The history of $\beta$-blockers has been heavily influenced by the finding of three different $\beta$-ARs subtypes that paved the way for the design and the development of selective agonists and antagonists with specific therapeutic goals. It is known that $\beta$1-ARs are mainly expressed in the heart, $\beta$2-ARs in the smooth muscle of vasculature and airways and $\beta$3-ARs in the adipose tissue [2]. Cardiovascular therapies were deeply influenced and revolutionized when propranolol was introduced for the first time in the clinic for its ability to reduce myocardial oxygen demand under angina episodes [5,6]. Since then, three generations of $\beta$-ARs antagonists have been developed, namely nonselective, $\beta$1-cardioselective and $\beta$1-blockers with vasodilation activity [2].

In the heart, $\beta$1-ARs, $\beta$2-ARs and $\beta$3-ARs are all expressed and modulate cardiac performance by activating key intracellular pathways [7]. The homeostatic role of $\beta$-ARs in cardiac physiology is significantly highlighted by their strong phylogenetic conservation among species [8–10].

However, the effects on the myocardium are predominantly mediated by $\beta$1-ARs and $\beta$2-ARs, since in the healthy heart they are more expressed with respect to $\beta$3-ARs [4,11].

Stress stimuli, such as increased activity of the SNS, cause augmented plasma catecholamine levels by an adaptive physiological response. However, if stress and high catecholamine levels become chronic, the overactivation of $\beta$-ARs may result in an aberrant signaling network that leads to cardiac remodeling, fibrosis, arrhythmia, ventricular hypertrophy and heart failure [4,11]. In this perspective, the role of $\beta$-blockers to counteract detrimental effects of chronic stress assumes a critical significance.

In terms of pharmacological classification, all $\beta$-blockers share a common feature that consists in their affinity/ability to bind $\beta$-ARs without evoking physiological effects. They are generally classified as competitive antagonists, so their effect can be overcome by increasing agonist concentration [12]. In particular, the blockade of cardiac $\beta$1-ARs is beneficial in myocardial hypertrophy, ischemia and failure, reducing both oxygen demand and renin production at a systemic level [13].

Much experimental and clinical evidence demonstrate that SNS overactivation and the resulting increase in catecholamine plasma levels is involved in the development of cardiac hypertrophy. In particular, the stimulation of $\beta$1-ARs, that in turn increases intracellular levels of cyclic-AMP and consequently of $Ca^{2+}$, may account for the activation of the cardiac hypertrophic pathway [14,15]. In this view, the possibility to design and develop even more selective $\beta$1-ARs blockers still represents an important goal for cardiovascular translational research and an open field to be investigated [16–20].

In the literature, several compounds endowed with a carbazole structure or with a simplified carbazole moiety are reported as $\beta$-blockers. In particular, carvedilol has been widely studied as a third-generation $\beta$-blocker [21], carazolol as a high affinity inverse agonist of the $\beta$-adrenergic receptor [22], pindolol as a nonselective $\beta$-blocker, which is used in the treatment of hypertension [23], and bucindolol (Figure 1) as a nonselective $\beta$-blocker with weak $\alpha$-blocking properties and intrinsic sympathomimetic activity [24].

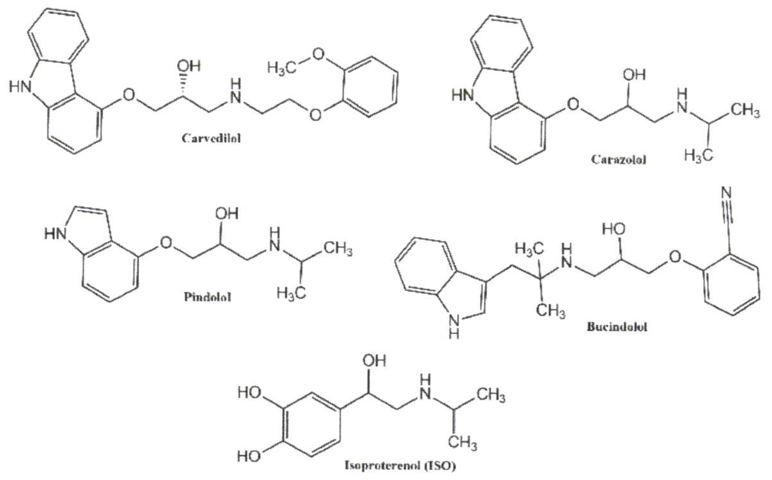

**Figure 1.** Representative structures of known $\beta$-blockers.

Based on this knowledge, the present study was undertaken to investigate the molecular interactions of a carbazole derivative, 4, and its three simplified analogs (7, 8, 10) with the active binding site of $\beta$1-AR. Considering the promising results obtained in silico, the in vitro bioactivity of these compounds was evaluated. In particular, we used the $\beta$-AR agonist isoproterenol (ISO, Figure 1) as a suitable hypertrophy-causing drug in H9c2 cells. This is a cell line widely used as model of cardiomyocytes for its biochemical, morphological and electrical/hormonal signaling properties [15]. Interesting results have been obtained for compounds **8** and **10** (Figure 2) making them novel tools against ISO-dependent in vitro cardiac hypertrophy.

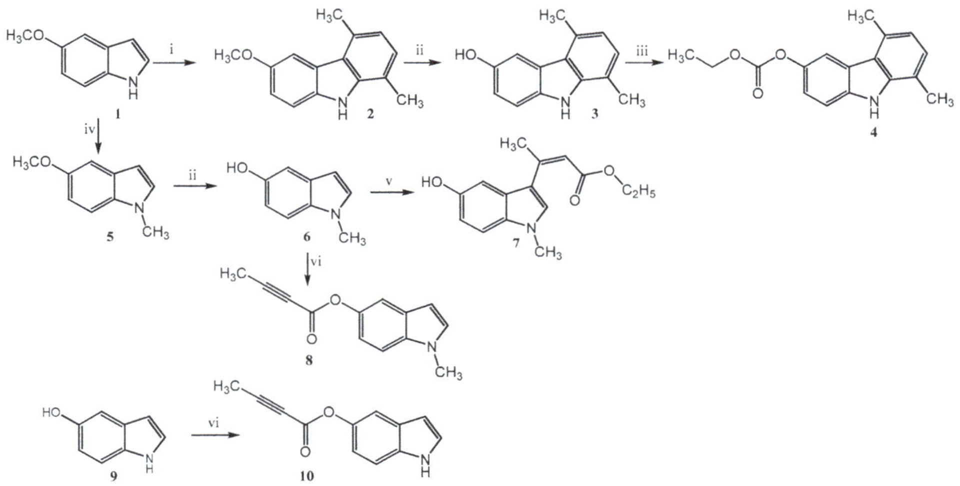

**Figure 2.** Reagents: (i) acetonylacetone, *p*-TSA, ethanol, reflux, 6 h; (ii) anhydrous pyridine hydrochloride, reflux, 2 h; (iii) ethyl chloroformate, NaOH solution 1N, acetone, reflux, 4 h; (iv) NaH, CH$_3$I, DME, 25 °C; (v) ethyl acetoacetate (1 mol), indium (III) chloride, reflux, 2 h; (vi) but-2-ynoic acid, DCC, DMAP, CH$_2$Cl$_2$/DMF (10:1), r.t. 2 h.

## 2. Materials and Methods

### 2.1. Chemistry

Commercial reagents were purchased from Aldrich, Acros Organics and Alfa Aesar and used without additional purification. Melting points were determined on a Kofler melting point apparatus. IR spectra were taken with a Perkin Elmer BX FT-IR. Mass spectra were taken on a JEOL JMS GCMate spectrometer at an ionizing potential of 70 eV (EI) or were performed using a spectrometer LC-MS Waters alliance 2695 (ESI+) or ESI mass spectrometer Finnigan LCQ Advantage max. $^1$H-NMR (400 MHz) and $^{13}$C-NMR (100 MHz) were recorded on a JEOL Lambda 400 Spectrometer. Chemical shifts were expressed in parts per million downfield from tetramethylsilane as an internal standard. The 2, 3, 5, 6 intermediates and the final product 4, 7, 8, 10 were prepared as described in the literature [25–30].

#### 2.1.1. 1,4-Dimethyl-6-methoxy-carbazole (**2**)

A mixture of 5-methoxyindole (**1**) (25.0 mmol), acetonylacetone (31.0 mmol) and *p*-TSA (20.0 mmol), in ethanol (60 mL) was heated to reflux for 6 h and then concentrated in vacuo. The crude product was purified by chromatography on silica gel using ethyl acetate:n-hexane, 2:8 as eluent. White solid, yield 57%, mp 150 °C. IR (KBr) (cm$^{-1}$): 3406, 2959, 1481, 1210, 1045, 812, 545. $^1$H NMR (DMSO-d$_6$): $\delta$ 11.01 (br, 1H, NH); 7.64 (s, 1H, Ar); 7.48 (d, J = 8.56 Hz, 1H, Ar); 7.09 (d, J = 8.28 Hz, 2H, Ar); 6.85 (d, J = 7.08 Hz, 1H, Ar); 3.90 (s, 3H, OCH$_3$); 2.69 (s, 3H, CH$_3$); 2.62 (s, 3H, CH$_3$). $^{13}$C NMR (DMSO-d$_6$): $\delta$ 152.5; 139.5;

134.4; 129.3; 125.3, 123.4; 120.1; 119.2; 117.1; 111.1; 105.0; 55.3; 19.9; 16.5. MS (ESI+): 226 (M$^+$ +1), 224 (M$^+$ −1).

### 2.1.2. 1,4-Dimethyl-6-hydroxy-carbazole (3)

A mixture of 1,4-dimethyl-6-methoxy-carbazole (**2**) and anhydrous pyridine hydrochloride (1:16) was heated to reflux for 2 h. The reaction mixture was left to cool to room temperature, then ice water was added. The product was extracted with Et$_2$O. The organic layer was washed with a solution of HCl (2N), dried over MgSO$_4$ and concentrated in vacuum. White solid, yield 75%, mp 174 °C. IR (KBr) (cm$^{-1}$): 3517, 3415, 1461, 1165, 847, 809, 543. $^1$H NMR (DMSO-d$_6$): δ 10.88 (s, 1H, NH); 9.12 (bs, 1H, OH); 7.53 (s, 1H, Ar); 7.40 (d, J = 8.56 Hz, 1H, Ar); 7.08 (d, J = 7.36 Hz, 1H, Ar); 6.95 (d, J = 8.56 Hz, 1H, Ar); 6.83 (d, J = 7.36 Hz, 1H, Ar); 2.61 (s, 3H, CH$_3$); 2.50 (s, 3H, CH$_3$). $^{13}$C NMR (DMSO-d$_6$): δ 150.2; 139.7; 133.8; 129.4; 125.4, 123.9; 120.3; 119.2; 117.2; 114.0; 111.1; 107.0; 20.1; 16.7. MS (ESI+): 212 (M$^+$ +1).

### 2.1.3. 5,8-Dimethyl-9H-carbazol-3-yl Ethyl Carbonate (4)

A mixture of NaOH solution 1N (50 mL) and acetone (50 mL) was added 1,4-dimethyl-6-hydroxy-carbazole (**3**) (31.0 mmol) and ethyl chloroformate (37.0 mmol). The reaction mixture was heated to reflux for 4 h and then was extracted with Et$_2$O. The organic layer was washed with a solution of NaOH (0.5%), dried over MgSO$_4$ and concentrated in vacuum. White solid, yield 91%, mp 130 °C. IR (KBr) (cm$^{-1}$): 3396, 1744, 1462, 1253, 1181, 1002, 811, 780, 552. $^1$H NMR (DMSO-d$_6$): δ 11.36 (s, 1H, NH); 7.94 (s, 1H, Ar); 7.58 (d, 1H, J = 8.56 Hz, Ar); 7.29 (dd, 1H, J$_1$ = 1.72 Hz, J$_2$ = 8.56 Hz, Ar); 7.16 (d, 1H, J = 7.32 Hz, Ar); 6.91 (d, 1H, J = 7.08 Hz, Ar); 4.35–4.28 (q, 2H, CH$_2$CH$_3$); 2.78 (s, 3H, CH$_3$); 2.57 (s, 3H, CH$_3$); 1.40–1.32 (t, 3H, CH$_2$CH$_3$). $^{13}$C NMR (DMSO-d$_6$): δ 153.9; 143.5; 139.8; 137.5; 129.7; 126.2, 123.1; 120.2; 120.0; 118.3; 117.6; 114.1; 111.0; 64.3; 20.0; 16.6; 14.0. MS (ESI+): 284 (M$^+$ +1). ESI *m/z* calcd for C$_{17}$H$_{18}$NO$_3$: 284.13; found: 284.10.

### 2.1.4. 5-Methoxy-1-methylindole (5)

To a stirred cold solution (0 °C) of 5-methoxyindole (**1**) (6.80 mmol) in dry DMF (30 mL), was added NaH 60% oil dispersion (10.20 mmol). After 10 min stirring at this temperature, iodomethane (20.40 mmol) was added and the mixture was further stirred at 25 °C for 1 h. Water (100 mL) was then added to the reaction mixture and the solid product obtained was filtered, washed with water and dried. White solid, yield 92%, mp 114 °C. IR (KBr) (cm$^{-1}$): 2922, 1621, 1496, 1242, 1151, 1025, 802, 725. $^1$H-NMR (DMSO-d$_6$) δ 7.30 (d, 1H, J = 8.80 Hz, Ar); 7.23 (d, 1H, J = 2.90 Hz, Ar); 7.04–7.01 (m, 1H, Ar); 6.76 (dd, 1H, J = 1.90, 8.80 Hz, Ar); 6.30 (d, 1H, J = 2.90 Hz Ar); 3.72 (s, 6H, NCH$_3$ and OCH$_3$). $^{13}$C NMR (DMSO-d$_6$): δ 154.0; 130.6; 129.0; 128.4; 112.3; 111.9; 111.6; 101.1; 56.0; 32.9. MS (ESI+): 162 (M$^+$ +1).

### 2.1.5. 5-Hydroxy-1-methylindole (6)

A mixture of 5-methoxy-1-methylindole (**5**) and anhydrous pyridine hydrochloride (1:16) was heated to reflux for 2 h. The reaction mixture was left to cool to 25 °C, then ice water was added. The product was extracted with Et$_2$O. The organic layer was washed with a solution of HCl (2N), dried over Na$_2$SO$_4$ and concentrated in vacuum. Yellow solid, yield 60%, mp 156 °C, IR (KBr) (cm$^{-1}$): 3177, 2924, 1621, 1489, 1234, 1145, 949, 795, 719. $^1$H-NMR (DMSO-d$_6$) δ 8.67 (s, 1H, OH); 7.19 (s, 1H, Ar); 7.17–7.15 (m, 1H, Ar); 6.81 (d, 1H, J = 3.00 Hz, Ar); 6.62 (dd, 1H, J = 1.90, 8.80 Hz, Ar); 6.18 (d, 1H, J = 2.90 Hz, Ar); 3.04 (s, 3H, NCH$_3$). $^{13}$C NMR (DMSO-d$_6$): δ 152.5; 130.1; 129.8; 128.7; 116.3; 110.1; 101.6; 100.1; 33.9. MS (ESI+): 148 (M$^+$ +1).

### 2.1.6. Ethyl 3-(5-hydroxy-1-methyl-1H-indol-3-yl)but-2-enoate (7)

To a mixture of 5-hydroxy-1-methyl-1*H*-indole (**6**) and ethyl acetoacetate (1 mol), indium(III) chloride (10 mol%) was added under nitrogen. The reaction mixture was heated under reflux for 2 h, and then it was left to cool to 25 °C. Ice water was added and then the reaction mixture was extracted by ethyl acetate. The organic layer was collected and washed with brine, dried over $Na_2SO_4$ and concentrated in vacuum. The solid residue was washed with $Et_2O$. Pink solid, yield 58%, mp 200 °C. IR (KBr) (cm$^{-1}$): 3400, 2931, 1705, 1488, 1373, 1329, 1201, 1083, 1016, 851, 788. $^1$H NMR (DMSO- $d_6$) δ 8.32 (s, 1H, Ar); 7.08–7.06 (m, 2H, Ar, OH); 6.52–6.49 (m, 2H, Ar); 6.45 (s, 1H, C=CH); 3.78–3.74 (q, 2H, $CH_2$); 3.67 (s, 3H, $NCH_3$); 1.83 (s, 3H, C-$CH_3$); 0.82–0.78 (t, 3H, $CH_3$). $^{13}$C NMR (DMSO-$d_6$): δ 200.4; 153.3; 152.5; 132.1; 129.4; 128.6; 125.1; 116.9; 109.1; 108.7; 103.6; 34.1; 33.9; 20.5; 8.7. MS (EI) $m/z$: 259 (M$^+$, 1). ESI $m/z$ calcd for $C_{15}H_{18}NO_3$: 260.13; found: 260.10.

### 2.1.7. 1-Methyl-1H-indol-5-yl-but-2-ynoate (**8**)

But-2-ynoic acid (1.1 eq), DCC (1 eq) and DMAP (cat) were added to a solution of 5-hydroxy-1-methyl-1H-indole (**6**) in a mixture of $CH_2Cl_2$ and DMF (10:1). The reaction mixture was stirred at 25 °C for 2 h and then concentrated in vacuum. The solid residue was suspended in ethyl acetate, filtered and the filtrate was washed with brine. The organic layer was dried over anhydrous $Na_2SO_4$ and evaporated under reduced pressure. The solid residue was purified by silica gel chromatography using ethyl acetate:n-hexane, 2:8 as an eluent.

White solid, yield 64%, mp 90 °C; IR (KBr) (cm$^{-1}$): 3402, 2233, 1711, 1625, 1487, 1246, 1217, 1123, 1033, 882. $^1$H NMR (CDCl$_3$) δ 7.28 (s, 1H, Ar); 7.23–7.19 (t, 1H, Ar); 7.01 (d, 1H, J = 3.00 Hz, Ar); 6.98 (dd, 1H, J = 2.00, 8.80 Hz, Ar); 6.46 (d, 1H, J = 3.00 Hz, Ar); 3.70 (s, 3H, $NCH_3$); 2.06 (s, 3H, C-$CH_3$). $^{13}$C NMR (CDCl$_3$): δ 149.4; 143.3; 133.5; 128.9; 128.4; 119.6; 109.4, 108.7; 100.1; 86.9; 76.1; 34.7; 3.6. MS (ESI+): 214 (M$^+$ +1). ESI $m/z$ calcd for $C_{13}H_{12}NO_2$: 214.09; found: 214.03.

### 2.1.8. Indol-5-yl-but-2-ynoate (**10**)

But-2-ynoic acid (1.1 eq), DCC (1 eq) and DMAP (cat) were added to a solution of 5-hydroxyindole (**9**) in a mixture of $CH_2Cl_2$ and DMF (10:1). The reaction mixture was stirred at room temperature for 2 h, and then concentrated in vacuum. The solid residue was suspended in ethyl acetate, filtered and the filtrate was washed with brine. The organic layer was dried over anhydrous $Na_2SO_4$ and evaporated under reduced pressure. The solid residue was purified by silica gel chromatography using ethyl acetate:petroleum ether, 2:8 as eluent.

White solid, yield 62%, mp 108 °C. IR (KBr) (cm$^{-1}$): 3410, 2237, 1705, 1480, 1261, 1121, 884. $^1$H NMR (CDCl$_3$) δ 8.25 (bs, 1H, NH); 7.40 (s, 1H, Ar); 7.20–7.23 (m, 1H, Ar); 7.03–7.01 (m, 1H, Ar); 6.95 (dd, 1H, J = 1.90, 8.80 Hz, Ar); 6.60–6.50 (m, 1H, Ar); 2.10 (s, 3H, C-$CH_3$). $^{13}$C NMR (CDCl$_3$): δ 149.1; 140.3; 134.5; 128.7; 124.4; 119.6; 111.4, 109.7; 102.1; 86.9; 75.1; 3.4. GC-MS $m/z$: 199 [M$^+$]. ESI $m/z$ calcd for $C_{12}H_9NO_2$: 199.06; found: 199.15.

### 2.2. Molecular Docking

Molecular docking of compounds 4, 7, 8 and 10 was carried out on the crystallographic structure of human $\beta$1-AR (PDB code 7BVQ). The binding pathway determines norepinephrine selectivity for the human beta 1 AR over beta 2 AR [22]. The molecular structures of the ligands were built by using the modeling software Avogadro [31]. Preliminary conversion of the structures from the PDB format was carried out by using the graphical interface AutoDock Tools 1.5.6 [32]. During the conversion, polar hydrogens were added for the crystallographic enzyme and apolar hydrogens of all compounds were merged to the carbon atom they were attached to. Docking calculations were performed by using AutoDock Vina1.1.2 [33] exploring the search volume that included the protein structure, and by performing a score-only assessment without any search in the case of redocking of the crystallographic ligand in its known binding mode. In the former case a very high exhaustiveness search was used, 8 time larger than the default value [34,35], facilitated by the relatively small number of active torsions around bond dihedral angles necessary to give full flexibility to the ligands (four degrees of freedom for 4 and three for

7, 8 and 10). The binding modes of the ligands were analyzed through visual inspection, and interactions energies and distances were quantified by using Molecular Operating Environment (MOE) 2018.01 (Chemical Computing Group ULC, Montreal, QC, Canada). The Molecular Graphics System PyMOL was used to visualize protein structure and the bonded ligands (PyMOL Molecular Graphics System, Version 1.2r3pre, Schrödinger, LLC., IN, New York, NY, USA)

*2.3. Cell Culture*

H9c2 cells (ATCC, CRL-1446, Manassas, VA, USA) and rat embryonic cardiomyocytes were cultured in DMEM/F-12 (Gibco, Thermo Fisher Scientific, Waltham, MA, USA) supplemented with 10% fetal bovine serum (FBS, Gibco) and 1% penicillin/streptomycin (Thermo Fisher Scientific) in a humidified incubator with 5% $CO_2$ at 37 °C. After reaching 80% confluence in 100-mm dishes, cells were detached by using 0.25% Trypsin-EDTA (1X) (Gibco) at 1:2 ratio following the manufacturer's instructions (ATCC). Cells were seeded and incubated for two days at 37 °C, 95% $O_2$ and 5% $CO_2$ before treatments.

*2.4. 3-(4,5-Dimethylthiazol-)2,5-diphenyl Tetrazolium Bromide (MTT) Assay*

H9c2 cell viability was evaluated by 3-(4,5-dimethylthiazol-2-yl)-2,5-diphenyltetrazolium bromide (MTT) assay, as previously described [36,37] Cells were seeded at a density of $5 \times 10^3$ cells/well in 96-well plate and then treated with compound 8 or compound 10 (1 nM–100 nM) for 24 h. Control cells were treated with vehicle. At the end of the treatments, the cell culture medium was replaced with 100 µL of 2 mg/mL MTT solution (Sigma Aldrich) and cells were incubated for 4 h at 37 °C, 5% $CO_2$. After MTT incubation, the solution was removed and the formazan crystals were solubilized in DMSO for 30 min. The absorbance was recorded using a Multiskan EX Microplate Reader Lab (Thermo Fisher Scientific) at 570 nm. The means of absorbance values of six wells in each experimental group were expressed as the percentage of cell viability. Cell viabilities were calculated as percentage of cell survival relative to the control [38,39].

*2.5. Morphological Analysis of H9c2 Cells*

The morphological alterations in H9c2 cells were evaluated by May-Grunwald Giemsa staining as previously described [36]. Cells were seeded in 60-mm dishes and exposed to the following treatments: control (vehicle), ISO (100 µM) alone, compound 8 (10, 25, 50, 75 and 100 nM), ISO + 8 (10, 25, 50, 75 and 100 nM), compound 10 (10, 25, 50, 75 and 100 nM), ISO + 10 (10, 25, 50, 75 and 100 nM) for 24 h. As positive control, the $\beta$-AR antagonist propranolol (PROP) (1 µM) was used alone or in cotreatment with ISO for 24 h [40]. After staining, cells were displayed by Olympus BX41 microscope and the images were taken with CSV1.14 software, using a CAM XC-30 for image acquisition. The cell surface area measurement was analyzed by using Image J 1.6 (NIH). Data were expressed as percentage of relative increase in cell surface area.

*2.6. Statistical Analysis*

Data, expressed as mean $\pm$ SEM, were analyzed by one-way ANOVA and Dunnett's Multiple Comparison Test (for post ANOVA comparisons) or nonparametric Newman-Keuls Multiple Comparison Test (for post-ANOVA comparisons) when appropriate. Values of * $p \leq 0.05$, ** $p \leq 0.01$, *** $p \leq 0.001$ were considered statistically significant versus the control group, while values of # $p \leq 0.05$, ## $p \leq 0.01$, ### $p \leq 0.001$ were considered statistically significant versus the ISO group. The statistical analysis was carried out using Graphpad Prism 5 (GraphPad Software, La Jolla, CA, USA).

## 3. Results and Discussion

*3.1. Chemistry*

Compound 4, endowed with a carbazole scaffold, and its simplified derivatives 7, 8, 10 were synthesized following chemical procedures previously reported (Figure 2) [25–28,41].

In particular, the carbazole 4 was prepared by a three-step synthesis. The starting material 5-methoxyndole 1, was used to obtain the 1,4-dimethyl-6-methoxy-carbazole (**2**) by Cranwell and Saxton reaction [26,42]. The demethylation of derivative **2**, by heating under reflux in anhydrous pyridine hydrochloride, gave the 1,4-dimethyl-6-hydroxy-carbazole (**3**) [27]. In the last step, the derivative but-2-ynoic acid (1.1 eq), DCC (1 eq) and DMAP (cat) were added. was prepared by reaction of 3 with ethyl chloroformate under standard conditions [43,44]. Ethyl 3-(5-hydroxy-1-metyl-1*H*-indol-3-yl)but-2-enoate (**7**) was instead prepared by von Pechmann reaction [41] using 5-hydroxy-1-methylindole (**6**) and Indium(III) chloride as catalyst. This latter, 6, was obtained starting from 5 by selective demethylation of the methoxyl group, in a similar manner as above reported for the preparation of derivative 3 [45]. 5-Methoxy-1-methylindole (**5**) was prepared using 1 as starting material 1 by *N*-methylation with NaH and methyl iodide. Finally,1-methyl-1*H*-indol-5-yl-but-2-ynoate (**8**) was synthesized by esterification of 6 with but-2-ynoic acid in a mixture of $CH_2Cl_2$, DMF, DCC and DMAP [41]. The same synthetic procedure was used for the preparation of indol-5-yl-but-2-ynoate (**10**) starting from 5-hydroxy-indole (**9**) [41].

*3.2. Molecular Docking*

In order to investigate the molecular interactions between compounds 4, 7, 8 and 10 and the active binding site of $\beta$1-adrenergic receptor, molecular docking studies were performed on a crystallographic structure of the target protein obtained from the Protein Data Bank (PDB code 7BVQ) [22]. In this structure the receptor is complexed with carazolol, a previously identified inverse agonist endowed with a carbazole based chemical structure [22]. In the PDB is reported another interesting crystallographic structure of $\beta$1 receptor (PDB code 7JJO), very recently deposited, in which the protein is complexed with ISO, a known $\beta$1-adrenergic receptor ligand. Even though this structure presents a more complete aminoacidic sequence, it does not belong to *Homo sapiens*. In spite of this shortcoming, it was compared after alignment with 7BVQ. From this comparison a sequence similarity of 86% was found between the two proteins and that an Asp residue in the protein binding site plays a key role during ligand-receptor interaction. In fact, in 7JJO ISO interacts with Asp121 forming a hydrogen bond and an ionic interaction through its nitrogen group, and with Val122 by a hydrophobic interaction through the aromatic moiety, whereas in 7BVQ the crystallographic ligand carazolol interacts by two hydrogen bonds with Asp1138 (corresponding to Asp121 of 7JJO). Beside this interaction, carazolol is anchored to Asn1363 of $\beta$1 receptor through its hydroxyl group by an additional hydrogen bond. The binding energy value obtained after redocking of carazolol was -7.7 kcal/mol. As shown in Figure 3, in our experiments, although the four molecules might adopt different orientations in the complex, all of them accommodate in the binding site and are able to interact with the key residues of the active site, including Asp1138 (Table 1).

**Table 1.** Binding Energy for compounds 4, 7, 8 and 10 in the $\beta$1-adrenergic receptor (PDB code 7BVQ).

| Compound | Binding Energy (kal/mol) |
|:---:|:---:|
| 4 | −8.0 |
| 7 | −7.7 |
| 8 | −7.5 |
| 10 | −7.8 |

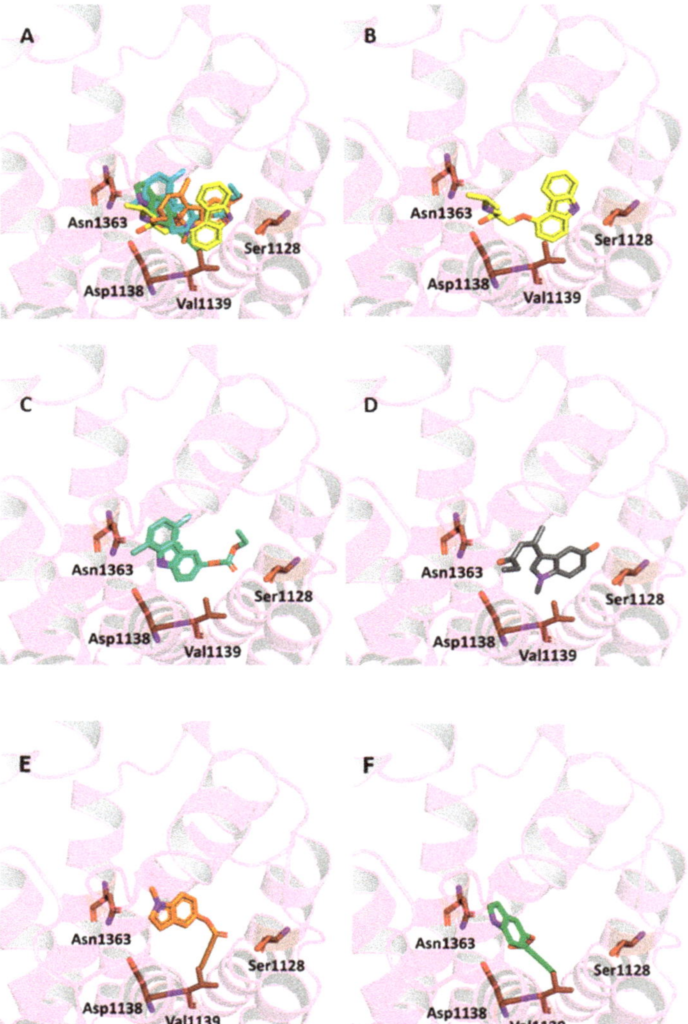

**Figure 3.** Ligand-binding pocket of the active site of the $\beta$1 adrenergic receptor. The protein backbone is represented in background as ribbons, and the key residues for ligand interaction (Asp1138, Asn1363, Ser1128, Val1139) are also indicated. (**A**) Superimposed binding modes of all the five ligands: carazolol (yellow), **4** (cyan), **7** (grey), **8** (orange) and **10** (green). Specific binding mode of carazolol (**B**), **4** (**C**), **7** (**D**), **8** (**E**) and **10** (**F**).

In particular, compound 4 interacts with the protein active site forming two hydrogen bonds: a first with Asp1138 through its carbazole nitrogen, with a distance of 3.15 Å, and the second with Ser1228 through its carbonyl group, with a distance of 2.84 Å, whereas compound 7 interacts by a hydrogen bond through its carbonyl group with Asp1138 with a distance of 3.43 Å. Compounds 8 and 10 although allocated in the same binding pocket interact through their side chain with Val1139 by hydrophobic interactions that stabilize the complexes. Overall, these results suggest that our compounds are able to accommodate in the active site of the $\beta$1 receptor with an affinity similar to that of the crystallographic ligands (Table 1).

### 3.3. In Vitro Findings

The interesting results obtained by in silico studies prompted us to investigate the biological activity exerted by compounds 4, 7, 8 and 10. Preliminary analyses indicated that compounds 4 and 7 exerted marked antiadrenergic effects on the Langendorff perfused heart model, reflecting a reduction of the cardiac performance associated with a weak cardiac recovery after washing with Krebs-Henseleit solution (unpublished data). Therefore, we aimed to deeper examine the biological activity induced by compounds 8 and 10. These molecules were tested on H9c2 cells alone or in the presence of ISO, widely used as inducer of myocardial stress and hypertrophy, mimicking the activation of the adrenergic and neurohumoral systems [46].

Effect of Compounds 8 and 10 on Cell Proliferation and ISO-Dependent Hypertrophy in H9c2 Cardiomyocytes

It has widely ascertained that the activation of $\beta 1$- and $\beta 2$-AR represents a main mechanism responsible for the progression of cardiac remodeling [47]. Although it has been shown that the expression of $\beta 2$-ARs is higher compared to that of $\beta 1$-ARs in both neonatal rat cardiomyocytes ($\beta 1$, 36% vs. $\beta 2$, 64%) [48] and in embryonic rat cardiomyocytes H9c2 cell lines [48,49], significant findings elucidated the selective contribution of $\beta 1$ and $\beta 2$-AR into the complex mechanisms of $\beta$-ARs-mediated adverse remodeling. Several studies also support the fact that, among the $\beta$-ARs, the $\beta 1$ subtype is mainly responsible for catecholamine effects and represents the main promoter of cardiomyocyte hypertrophy, interstitial fibrosis and heart failure [50,51]. For instance, it has been reported that isoproterenol binds to $\beta 1$-AR with higher affinity compared to $\beta 2$-AR [52,53]. Furthermore, the isoproterenol-mediated cardiac responses were inhibited by betaxolol, a $\beta 1$-AR antagonist, but not by ICI 118551, a $\beta 2$-AR antagonist. These effects were observed both in vitro (i.e., primary cultures of cardiac ventricular myocytes stimulated with isoproterenol) and in vivo (i.e., rats exposed to continuous infusion of isoproterenol) [48]. Overall, these observations strongly suggest that the $\beta 1$-subtype predominantly mediates isoproterenol-dependent hypertrophy in the heart.

Accordingly, to assess the effects of compounds 8 and 10 on cell viability, we first exposed H9c2 cells to increasing concentrations (1 to 100 nM) of each compound and their cytotoxic potential was evaluated by MTT assay. As shown in Figure 4A, compound 8 did not exert significant effects on cell proliferation at any tested concentration compared to the control cells, while compound 10 induced a significant increase in cardiac cell viability at 10 nM and 50 nM (Figure 4B). These data indicate that both compounds do not negatively affect cell viability and do not show direct cytotoxic effects.

Taking into consideration these results, we evaluated the in vitro effects of compounds 8 and 10 against cardiomyocytes hypertrophy. Therefore, we exposed H9c2 cells to ISO at 100 $\mu$M for 24 h [36], in the presence or absence of compound 8 or compound 10. Morphological staining analysis revealed that ISO treatment induced a significant increase in cardiac cell size with respect to control cells, indicating that the model of in vitro hypertrophy was successfully established (Figure 5A). Notably, compound 8 resulted was effective in significantly reducing cell size compared to the ISO group only at concentrations of 25 nM and 100 nM. Cells treated with compound 8 alone did not show significant increase in cell size (Figure 5A). The surface area analysis of H9c2 cells reflected the same trend (Figure 5B).

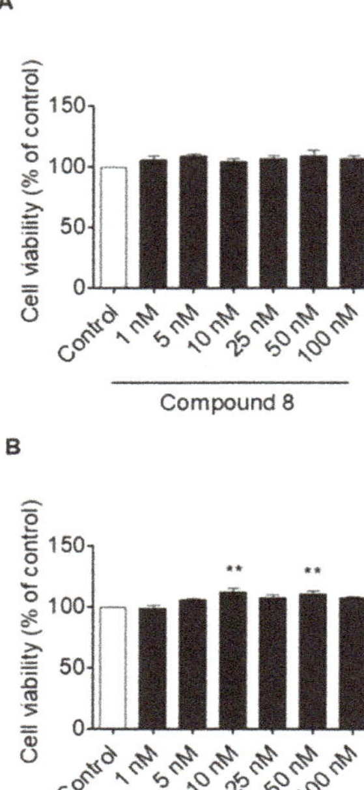



**Figure 4.** Effects of compounds 8 and 10 on cell viability in H9c2 cardiomyocytes. H9c2 cells were treated with vehicle (Control) or increasing concentrations of (**A**) compound **8** (1 nM–100 nM) or (**B**) compound **10** (1 nM–100 nM) for 24 h. Cell viability was determined using the MTT assay and was expressed as the percentage of control cells exposed only to vehicle (indicated as Control). Results are represented as mean ± SEM ($n$ = 6 per group). Significant differences were detected by one-way ANOVA followed by Dunnett's test, $p < 0.01$ (**) vs. Control group.

The same experiment was carried out for testing the activity of compound 10. In this case, as evinced by morphological staining and cell surface area analysis shown in Figure 6, ISO-treated cells exhibited a significant cardiac cell size increase compared to the control cells. Conversely, in the ISO + compound 10 (10 nM) treated group, cell size was significantly decreased compared to those treated with ISO alone. Compound 10 was unable to counteract in a significant manner the effects of ISO at higher doses (25 to 100 nM). These findings indicate that, similar to compound 8, compound 10 was able to counteract ISO-dependent effects on H9c2 cells. However, the first effective dose of compound 10 was lower than that of compound 8 (10 nM vs. 25 nM). Figure 6 also indicates that compound 10 alone did not exert significant effects on cell size at 10 nM, while this parameter was significantly increased when administered at higher doses (i.e., 50 nM and 75 nM).

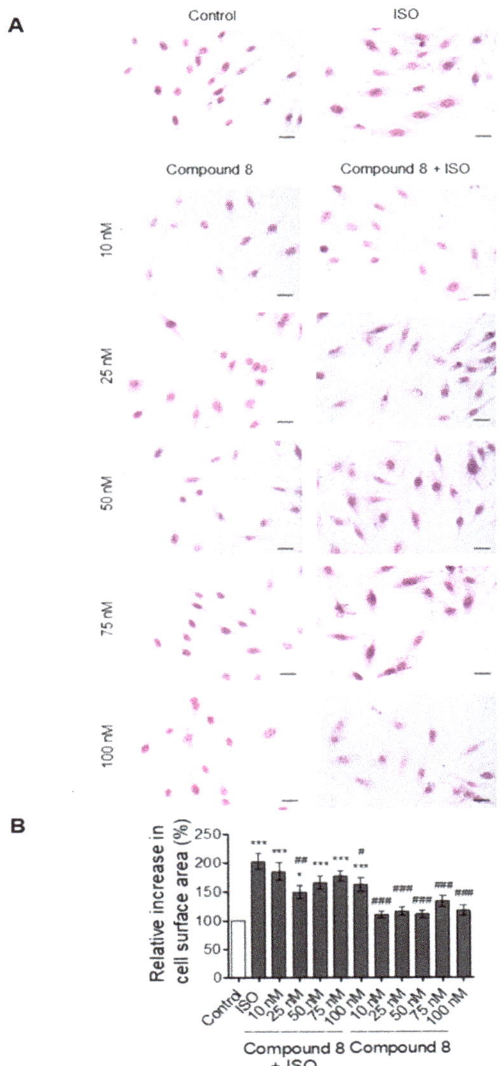


**Figure 5.** Effects of compound 8 on cardiomyocytes hypertrophy induced by isoproterenol (ISO) in vitro. (**A**) Morphological staining of H9c2 cells exposed to vehicle (indicated as Control), ISO (100 μM), ISO + compound 8 (from 10 nM to 100 nM) and compound **8** (from 10 nM to 100 nM) for 24 h. Scale bars: 25 μm. (**B**) Cell surface area (%) of H9c2 cells exposed to vehicle (indicated as Control), ISO (100 μM), ISO + compound 8 (from 10 nM to 100 nM) and compound 8 (from 10 nM to 100 nM) for 24 h. Data are expressed as mean ± SEM derived from two independent measurements for each group. $p < 0.05$ (*); $p < 0.001$ (***) vs. Control group; $p < 0.05$ (#); $p < 0.01$ (##); $p < 0.001$ (###) vs. ISO group (one-way ANOVA and Newman-Keuls multiple comparison test).

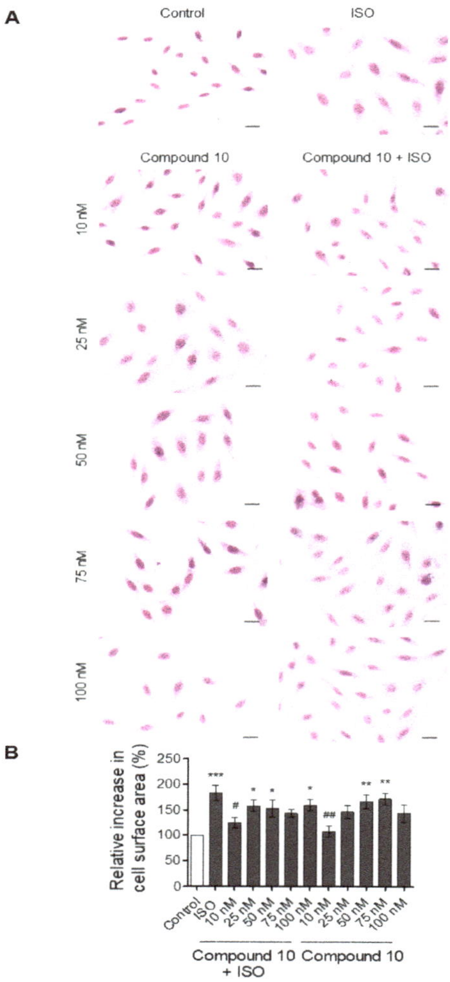



**Figure 6.** Effects of compound 10 on cardiomyocytes hypertrophy induced by isoproterenol (ISO) in vitro. (**A**) Morphological staining of H9c2 cells exposed to vehicle (indicated as Control), ISO (100 μM), ISO + compound 10 (from 10 nM to 100 nM) and compound 10 (from 10 nM to 100 nM) for 24 h. Scale bars: 25 μm. (**B**) Cell surface area (%) of H9c2 cells exposed to vehicle (indicated as Control), ISO (100 μM), ISO + compound 10 (from 10 nM to 100 nM) and compound 10 (from 10 nM to 100 nM) for 24 h. Data are expressed as mean ± SEM derived from two independent measurements for each group. $p < 0.05$ (*); $p < 0.01$ (**); $p < 0.001$ (***) vs. Control group; $p < 0.05$ (#); $p < 0.01$ (##) vs. ISO group (one-way ANOVA and Newman-Keuls multiple comparison test).

Our data are of interest in the view of $\beta$-blocker monitoring and dosing, to avoid or minimize their side effects [54]. $\beta$-blockers reach their therapeutic goal mainly by inducing negative chronotropic effects that prolong heart diastole and rest. In this context the choice of an appropriate dose is vital to avoid concomitant adverse negative inotropic effects.

Accordingly, we used the $\beta$-adrenergic receptor antagonist propranolol as a positive control. As depicted in Figure 7, both morphological staining and cell surface area indicated that in H9c2 cotreated with propranolol (1 μM), ISO failed to increase cardiomyocyte size.

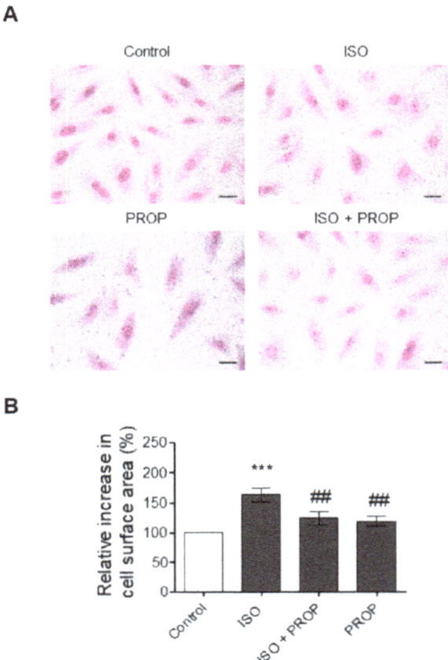


**Figure 7.** Effects of propranolol (PROP) on cardiomyocytes hypertrophy induced by isoproterenol (ISO) in vitro. (**A**) Morphological staining of H9c2 cells exposed to vehicle (indicated as Control), ISO (100 µM), ISO + propranolol (PROP) (1 µM) and propranolol (1 µM) for 24 h. Scale bars: 25 µm. (**B**) Cell surface area (%) of H9c2 cells exposed to vehicle (indicated as Control), ISO (100 µM), ISO + propranolol (PROP) (1 µM) and propranolol (1 µM) for 24 h. Data are expressed as mean ± SEM derived from two independent measurements for each group. $p < 0.001$ (***) vs. Control group; $p < 0.01$ (##) vs. ISO group (one-way ANOVA and Newman-Keuls multiple comparison test).

Our compounds were effective at lower doses (nM range) with respect to the classic β-blocker propranolol (µM range), according to data that support beneficial effects of low dosage especially in patients suffering from ventricular hypertrophy [55]. Clinical evidence indicates that, in practice, β-blockers are often used at half or lower doses compared to manufacturers' recommendations [56]. For some drugs, lower doses even improved the disease outcomes [57]. In the specific case of β-blockers, much data indicate their best use at the lowest effective dose [58].

## 4. Conclusions

In summary, our in silico results support the hypothesis that compounds 4, 7, 8 and 10 could interact with the active binding site of β1-AR. Moreover, compounds 8 and 10 emerged as potential counteracting agents against ISO-dependent in vitro cardiac hypertrophy. These data are of additional interest considering that the effect of the two molecules was achieved starting from lower concentrations compared to those of the traditional β-AR antagonist propranolol. In this context, our compounds could act as more potent pharmacological agents (i.e., β1-blockers), and their lower effective doses may contribute to minimize their adverse effects. Altogether, our findings depict the studied molecules worthy of further in vivo investigation.

**Author Contributions:** Conceptualization, A.C. (Anna Caruso) and T.A.; methodology, F.G., A.D.B.; validation, A.C. (Alessia Carocci), T.A.; formal analysis, V.R. and M.A.O.; investigation, T.P. and M.A.O.; writing—original draft preparation, F.G., A.C. (Anna Caruso) and C.R. writing—review and editing, A.C. (Anna Caruso) and M.S.S.; supervision, M.S.S. All authors have read and agreed to the published version of the manuscript.

**Funding:** This research received no external funding.

**Institutional Review Board Statement:** Not applicable.

**Informed Consent Statement:** Not applicable.

**Data Availability Statement:** Data are contained within the article.

**Acknowledgments:** We thank Andrea Brancale of Cardiff University for the use of molecular docking programs, including the MOE software package. M.A. Occhiuzzi also acknowledges kind hospitality in the School of Pharmacy and Pharmaceutical Science (Cardiff, Wales, UK). C.R. acknowledges POR Calabria (Italy) FESR-FSE 2014/2020- Azione 10.5.12-Linea B (DR n. 683 del 21/05/2019) for financial support for the RTDa position.

**Conflicts of Interest:** The authors declare no conflict of interest.

## References

1. Do Vale, G.T.; Ceron, C.S.; Gonzaga, N.A.; Simplicio, J.A.; Padovan, J.C. Three generations of β-blockers: History, class differences and clinical applicability. *Curr. Hypertens. Rev.* **2018**, *15*, 22–31. [CrossRef] [PubMed]
2. Oliver, E.; Mayor, F., Jr.; D'Ocon, P. Beta-blockers: Historical Perspective and mechanisms of action. *Rev. Española Cardiol.* **2019**, *72*, 853–862. [CrossRef]
3. Dale, H.H. On some physiological actions of ergot. *J. Physiol.* **1906**, *34*, 163–206. [CrossRef]
4. Wang, J.; Gareri, C.; Rockman, H.A. G-protein-coupled receptors in heart disease. *Circ. Res.* **2018**, *123*, 716–735. [CrossRef]
5. Moran, N.C.; Perkins, M.E. Adrenergic blockade of the mammalian heart by a dichloro analogue of isoproterenol. *J. Pharm. Exp. Ther.* **1958**, *124*, 223–237.
6. Quirke, V. Putting theory into practice: James Black, receptor theory and the development of the beta-blockers at ICI, 1958–1978. *Med. Hist.* **2006**, *50*, 69–92. [CrossRef] [PubMed]
7. Pfleger, J.; Gresham, K.; Koch, W.J. G protein-coupled receptor kinases as therapeutic targets in the heart. *Nat. Rev. Cardiol.* **2019**, *16*, 612–622. [CrossRef]
8. Giltrow, E.; Eccles, P.D.; Hutchinson, T.H.; Sumpter, J.P.; Rand-Weaver, M. Characterisation and expression of β1-, β2- and β3-adrenergic receptors in the fathead minnow (*Pimephales promelas*). *Gen. Comp. Endocrinol.* **2011**, *173*, 483–490. [CrossRef]
9. Leo, S.; Gattuso, A.; Mazza, R.; Filice, M.; Cerra, M.C.; Imbrogno, S. Cardiac influence of the β3-adrenoceptor in the goldfish (*Carassius auratus*): A protective role under hypoxia? *J. Exp. Biol.* **2019**, *222*, jeb211334. [CrossRef]
10. Imbrogno, S.; Filice, M.; Cerra, M.C. Exploring cardiac plasticity in teleost: The role of humoral modulation. *Gen. Comp. Endocrinol.* **2019**, *283*, 113236. [CrossRef]
11. De Lucia, C.; Eguchi, A.; Koch, W.J. New insights in cardiac β-Adrenergic signaling during heart failure and aging. *Front. Pharm.* **2018**, *9*, 904. [CrossRef]
12. Baker, J.G.; Hill, S.J.; Summers, R.J. Evolution of β-blockers: From anti-anginal drugs to ligand-directed signalling. *Trends Pharm. Sci.* **2011**, *32*, 227–234. [CrossRef]
13. Poirier, L.; Tobe, S.W. Contemporary use of β-blockers: Clinical relevance of subclassification. *Can. J. Cardiol.* **2014**, *30*, S9–S15. [CrossRef] [PubMed]
14. Han, J.W.; Kang, C.; Kim, Y.; Lee, M.G.; Kim, J.Y. Isoproterenol-induced hypertrophy of neonatal cardiac myocytes and H9c2 cell is dependent on TRPC3-regulated CaV1.2 expression. *Cell Calcium* **2020**, *92*, 102305. [CrossRef]
15. Hescheler, J.; Meyer, R.; Plant, S.; Krautwurst, D.; Rosenthal, W.; Schultz, G. Morphological, biochemical, and electrophysiological characterization of a clonal cell (H9c2) line from rat heart. *Circ. Res.* **1991**, *69*, 1476–1486. [CrossRef]
16. Angelone, T.; Caruso, A.; Rochais, C.; Caputo, A.M.; Cerra, M.C.; Dallemagne, P.; Filice, E.; Genest, D.; Pasqua, T.; Puoci, F.; et al. Indenopyrazole oxime ethers: Synthesis and β1-adrenergic blocking activity. *Eur. J. Med. Chem.* **2015**, *92*, 672–681. [CrossRef]
17. Grande, F.; Parisi, O.I.; Mordocco, R.A.; Rocca, C.; Puoci, F.; Scrivano, L.; Quintieri, A.M.; Cantafio, P.; Ferla, S.; Brancale, A.; et al. Quercetin derivatives as novel antihypertensive agents: Synthesis and physiological characterization. *Eur. J. Pharm. Sci.* **2016**, *82*, 161–170. [CrossRef] [PubMed]
18. Di Nunno, L.; Franchini, C.; Scilimati, A.; Sinicropi, M.S.; Tortorella, P. Chemical and chemoenzymatic routes to 1-(benzothiazol-2-ylsulfanyl)-3-chloropropan-2-ol, a precursor of drugs with potential β-blocker activity. *Tetrahedron Asymmetry* **2000**, *11*, 1571–1583. [CrossRef]
19. Go, J.G.; Santiago, L.D.; Miranda, A.C.; Jara, R.D. Effect of Beta-blockers on Hypertension and Heart Failure with Reduced Ejection Fraction: A Systematic Review of Randomized Controlled Trials. *Hypertens. J.* **2019**, *5*. [CrossRef]
20. Bennett, M.; Chang, C.L.; Tatley, M.; Savage, R.; Hancox, R.J. The safety of cardio-selective beta1-blockers in asthma: Literature review and search of global pharmacovigilance safety reports. *ERJ Open Res.* **2021**, *7*, 00801–02020. [CrossRef] [PubMed]
21. Chang, A.; Yeung, S.; Thakkar, A.; Huang, K.M.; Liu, M.M.; Kanassatega, R.-S.; Parsa, C.; Orlando, R.; Jackson, E.K.; Andresen, B.T.; et al. Prevention of skin carcinogenesis by the β-blocker carvedilol. *Prev. Cancer Prev. Res.* **2015**, *8*, 27–36. [CrossRef] [PubMed]
22. Xu, X.; Kaindl, J.; Clark, M.J.; Hübner, H.; Hirata, K.; Sunahara, R.K.; Gmeiner, P.; Kobilka, B.K.; Liu, X. Binding pathway determines norepinephrine selectivity for the human β1AR over β2AR. *Cell Res.* **2021**, *31*, 569–579. [CrossRef]

23. Wong, G.W.K.; Boyda, H.N.; Wright, J.M. Blood pressure lowering efficacy of partial agonist beta blocker monotherapy for primary hypertension. *Cochrane Database Syst. Rev.* **2014**, *12*, 2017. [CrossRef]
24. Willette, R.N.; Mitchell, M.P.; Ohlstein, E.H.; Lukas, M.A.; Ruffolo, R.R. Evaluation of intrinsic sympathomimetic activity of bucindolol and carvedilol in rat heart. *Pharmacology* **1998**, *56*, 30–36. [CrossRef] [PubMed]
25. Caruso, A.; Sophie Voisin-Chiret, A.; Lancelot, J.-C.; Sinicropi, S.; Garofalo, A.; Rault, S. Novel and efficient synthesis of 5,8-dimethyl-9h-carbazol-3-ol via a hydroxydeboronation reaction. *Heterocycles* **2007**, *71*, 2203–2210. [CrossRef]
26. Saturnino, C.; Grande, F.; Aquaro, S.; Caruso, A.; Iacopetta, D.; Bonomo, M.; Longo, P.; Schols, D.; Sinicropi, M. Chloro-1,4-dimethyl-9H-carbazole Derivatives Displaying Anti-HIV Activity. *Molecules* **2018**, *23*, 286. [CrossRef] [PubMed]
27. Rizza, P.; Pellegrino, M.; Caruso, A.; Iacopetta, D.; Sinicropi, M.S.; Rault, S.; Lancelot, J.C.; El-Kashef, H.; Lesnard, A.; Rochais, C.; et al. 3-(Dipropylamino)-5-hydroxybenzofuro[2,3-*f*]quinazolin-1(2*H*)-one (DPA-HBFQ-1) plays an inhibitory role on breast cancer cell growth and progression. *Eur. J. Med. Chem.* **2016**, *107*, 275–287. [CrossRef]
28. Sinicropi, M.S.; Iacopetta, D.; Rosano, C.; Randino, R.; Caruso, A.; Saturnino, C.; Muià, N.; Ceramella, J.; Puoci, F.; Rodriquez, M.; et al. N-thioalkylcarbazoles derivatives as new anti-proliferative agents: Synthesis, characterisation and molecular mechanism evaluation. *J. Enzym. Inhib. Med. Chem.* **2018**, *33*, 434–444. [CrossRef]
29. Caruso, A.; Lancelot, J.C.; El-Kashef, H.; Sinicropi, M.S.; Legay, R.; Lesnard, A.; Rault, S. A rapid and versatile synthesis of novel pyrimido[5,4-b]carbazoles. *Tetrahedron* **2009**, *65*, 10400–10405. [CrossRef]
30. Saturnino, C.; Caruso, A.; Longo, P.; Capasso, A.; Pingitore, A.; Caroleo, M.; Cione, E.; Perri, M.; Nicolo, F.; Nardo, V.; et al. Crystallographic Study and Biological Evaluation of 1,4-dimethyl-N-alkylcarbazoles. *Curr. Top. Med. Chem.* **2015**, *15*, 973–979. [CrossRef]
31. Hanwell, M.D.; Curtis, D.E.; Lonie, D.C.; Vandermeerschd, T.; Zurek, E.; Hutchison, G.R. Avogadro: An advanced semantic chemical editor, visualization, and analysis platform. *J. Cheminform.* **2012**, *4*, 17. [CrossRef]
32. Morris, G.M.; Goodsell, D.S.; Halliday, R.S.; Huey, R.; Hart, W.E.; Belew, R.K.; Olson, A.J. Automated docking using a Lamarckian genetic algorithm and an empirical binding free energy function. *J. Comput. Chem.* **1998**, *19*, 1639–1662. [CrossRef]
33. Trott, O.; Olson, A.J. AutoDock Vina: Improving the speed and accuracy of docking with a new scoring function, efficient optimization, and multithreading. *J. Comput. Chem.* **2010**, *31*, 455–461. [CrossRef]
34. Grande, F.; Rizzuti, B.; Occhiuzzi, M.A.; Ioele, G.; Casacchia, T.; Gelmini, F.; Guzzi, R.; Garofalo, A.; Statti, G. Identification by molecular docking of homoisoflavones from leopoldia comosa as ligands of estrogen receptors. *Molecules* **2018**, *23*, 894. [CrossRef]
35. Casacchia, T.; Occhiuzzi, M.A.; Grande, F.; Rizzuti, B.; Granieri, M.C.; Rocca, C.; Gattuso, A.; Garofalo, A.; Angelone, T.; Statti, G. A pilot study on the nutraceutical properties of the Citrus hybrid Tacle® as a dietary source of polyphenols for supplementation in metabolic disorders. *J. Funct. Foods* **2019**, *52*, 370–381. [CrossRef]
36. Rocca, C.; Grande, F.; Granieri, M.C.; Colombo, B.; De Bartolo, A.; Giordano, F.; Rago, V.; Amodio, N.; Tota, B.; Cerra, M.C.; et al. The chromogranin A$_{1-373}$ fragment reveals how a single change in the protein sequence exerts strong cardioregulatory effects by engaging neuropilin-1. *Acta Physiol.* **2020**, *231*, e13570.
37. Rocca, C.; Femminò, S.; Aquila, G.; Granieri, M.C.; De Francesco, E.M.; Pasqua, T.; Rigiracciolo, D.C.; Fortini, F.; Cerra, M.C.; Maggiolini, M.; et al. Notch1 mediates preconditioning protection induced by gper in normotensive and hypertensive female rat hearts. *Front. Physiol.* **2018**, *9*, 521. [CrossRef]
38. Nettore, I.C.; Rocca, C.; Mancino, G.; Albano, L.; Amelio, D.; Grande, F.; Puoci, F.; Pasqua, T.; Desiderio, S.; Mazza, R.; et al. Quercetin and its derivative Q2 modulate chromatin dynamics in adipogenesis and Q2 prevents obesity and metabolic disorders in rats. *J. Nutr. Biochem.* **2019**, *69*, 151–162. [CrossRef]
39. Rocca, C.; De Bartolo, A.; Grande, F.; Rizzuti, B.; Pasqua, T.; Giordano, F.; Granieri, M.C.; Occhiuzzi, M.A.; Garofalo, A.; Amodio, N.; et al. Cateslytin abrogates lipopolysaccharide-induced cardiomyocyte injury by reducing inflammation and oxidative stress through toll like receptor 4 interaction. *Int. Immunopharmacol.* **2021**, *94*, 107487. [CrossRef]
40. Cai, L.; Fan, G.; Wang, F.; Liu, S.; Li, T.; Cong, X.; Chun, J.; Chen, X. Protective role for LPA3 in cardiac hypertrophy induced by myocardial infarction but not by isoproterenol. *Front. Physiol.* **2017**, *8*, 356. [CrossRef]
41. Sinicropi, M.S.; Caruso, A.; Conforti, F.; Marrelli, M.; El Kashef, H.; Lancelot, J.; Rault, S.; Statti, G.A.; Menichini, F. Synthesis, inhibition of NO production and antiproliferative activities of some indole derivatives. *J. Enzyme Inhib. Med. Chem.* **2009**, *24*, 1148–1153. [CrossRef]
42. Cranwell, P.A.; Saxton, J.E. A synthesis of ellipticine. *J. Chem. Soc.* **1962**, *683*, 3482–3487. [CrossRef]
43. Letois, B.; Lancelot, J.; Rault, S.; Robba, M.; Tabka, T.; Gauduchon, P.; Bertreux, E.; Le Talaer, J. Étude de la cytotoxicitéin vitro de dérivés du carbazole III. 3-Amino et 3-nitro-1,4-diméthyl-9H-carbazoles diversement substitués en position. *Eur. J. Med. Chem.* **1990**, *25*, 775–784.
44. Yadav, M.R.; Grande, F.; Chouhan, B.S.; Naik, P.P.; Giridhar, R.; Garofalo, A.; Neamati, N. Cytotoxic potential of novel 6,7-dimethoxyquinazolines. *Eur. J. Med. Chem.* **2012**, *48*, 231–243. [CrossRef] [PubMed]
45. Caruso, A.; Chimento, A.; El-Kashef, H.; Lancelot, J.-C.; Panno, A.; Pezzi, V.; Saturnino, C.; Sinicropi, M.S.; Sirianni, R.; Rault, S. Antiproliferative activity of some 1,4-dimethylcarbazoles on cells that express estrogen receptors: Part I. *J. Enzyme Inhib. Med. Chem.* **2012**, *27*, 609–613. [CrossRef]
46. Lymperopoulos, A.; Rengo, G.; Koch, W.J. Adrenergic nervous system in heart failure: Pathophysiology and therapy. *Circ. Res.* **2013**, *113*, 739–753. [CrossRef] [PubMed]

47. Xu, Q.; Dalic, A.; Fang, L.; Kiriazis, H.; Ritchie, R.H.; Sim, K.; Gao, X.M.; Drummond, G.; Sarwar, M.; Zhang, Y.Y.; et al. Myocardial oxidative stress contributes to transgenic β 2- adrenoceptor activation-induced cardiomyopathy and heart failure. *Br. J. Pharm.* **2011**, *162*, 1012–1028. [CrossRef]

48. Morisco, C.; Zebrowski, D.C.; Vatner, D.E.; Vatner, S.F.; Sadoshima, J. β-adrenergic cardiac hypertrophy is mediated primarily by the β₁-subtype in the rat heart. *J. Mol. Cell. Cardiol.* **2001**, *33*, 561–573. [CrossRef]

49. Dangel, V.; Giray, J.; Ratge, D.; Wisser, H. Regulation of β-adrenoceptor density and mRNA levels in the rat heart cell-line H9c. *Biochem. J.* **1996**, *317*, 925–931. [CrossRef]

50. Lohse, M.J.; Engelhardt, S.; Eschenhagen, T. What is the role of β-adrenergic signaling in heart failure? *Circ. Res.* **2003**, *93*, 896–906. [CrossRef]

51. Vidal, M.; Wieland, T.; Lohse, M.J.; Lorenz, K. β-Adrenergic receptor stimulation causes cardiac hypertrophy via a Gβγ/Erk-dependent pathway. *Cardiovasc. Res.* **2012**, *96*, 255–264. [CrossRef]

52. Hoffmann, C.; Leitz, M.R.; Oberdorf-Maass, S.; Lohse, M.J.; Klotz, K.N. Comparative pharmacology of human β-adrenergic receptor subtypes-characterization of stably transfected receptors in CHO cells. *Naunyn. Schmiedebergs. Arch. Pharm.* **2004**, *369*, 151–159. [CrossRef] [PubMed]

53. Del Carmine, R.; Ambrosio, C.; Sbraccia, M.; Cotecchia, S.; Ijzerman, A.P.; Costa, T. Mutations inducing divergent shifts of constitutive activity reveal different modes of binding among catecholamine analogues to the β₂-adrenergic receptor. *Br. J. Pharm.* **2002**, *135*, 1715–1722. [CrossRef] [PubMed]

54. Basile, J.N. Titration of β-blockers in heart failure: How to maximize benefit while minimizing adverse events. *Postgrad Med.* **2003**, *113*, 63–70. [CrossRef]

55. Schmitt, B.; Li, T.; Kutty, S.; Khasheei, A.; Schmitt, K.R.L.; Anderson, R.H.; Lunkenheimer, P.P.; Berger, F.; Kühne, T.; Peters, B. Call for papers impact of sympathoexcitation on cardiovascular function in humans effects of incremental beta-blocker dosing on myocardial mechanics of the human left ventricle: MRI 3D-tagging insight into pharmacodynamics supports theory of inner antagonism. *Am. J. Physiol. Heart Circ. Physiol.* **2015**, *309*, H45–H52.

56. Gislason, G.H.; Rasmussen, J.N.; Abildstrøm, S.Z.; Gadsbøll, N.; Buch, P.; Friberg, J.; Rasmussen, S.; Køber, L.; Stender, S.; Madsen, M.; et al. Long-term compliance with beta-blockers, angiotensin-converting enzyme inhibitors, and statins after acute myocardial infarction. *Eur. Heart J.* **2006**, *27*, 1153–1158. [CrossRef] [PubMed]

57. Dimmitt, S.B.; Stampfer, H.G. Low drug doses may improve outcomes in chronic disease. *Med. J. Aust.* **2009**, *191*, 511–513. [CrossRef] [PubMed]

58. Dimmitt, S.B.; Stampfer, H.G.; Warren, J.B. β-adrenoceptor blockers valuable but higher doses not necessary. *Br. J. Clin. Pharm.* **2014**, *78*, 1076–1079. [CrossRef]

*Review*

# Carbazole Derivatives as STAT Inhibitors: An Overview

**Anna Caruso** [1,†], **Alexia Barbarossa** [2,†], **Alessia Carocci** [2], **Giovanni Salzano** [3], **Maria Stefania Sinicropi** [1,*] **and Carmela Saturnino** [3]

1   Department of Pharmacy, Health and Nutritional Sciences, University of Calabria, 87036 Arcavacata di Rende, Italy; anna.caruso@unical.it
2   Department of Pharmacy-Pharmaceutical Sciences, University of Bari Aldo Moro, 70126 Bari, Italy; alexia.barbarossa@uniba.it (A.B.); alessia.carocci@uniba.it (A.C.)
3   Department of Science, University of Basilicata, 85100 Potenza, Italy; giovanni.salzano@unibas.it (G.S.); carmela.saturnino@unibas.it (C.S.)
*   Correspondence: s.sinicropi@unical.it; Tel.: +39-0984-493019
†   These authors equally contributed to this work.

**Abstract:** The carbazole class is made up of heterocyclically structured compounds first isolated from coal tar. Their structural motif is preponderant in different synthetic materials and naturally occurring alkaloids extracted from the taxonomically related higher plants of the genus *Murraya*, *Glycosmis*, and *Clausena* from the *Rutaceae* family. Concerning the biological activity of these compounds, many research groups have assessed their antiproliferative action of carbazoles on different types of tumoral cells, such as breast, cervical, ovarian, hepatic, oral cavity, and small-cell lung cancer, and underlined their potential effects against psoriasis. One of the principal mechanisms likely involved in these effects is the ability of carbazoles to target the JAK/STATs pathway, considered essential for cell differentiation, proliferation, development, apoptosis, and inflammation. In this review, we report the studies carried out, over the years, useful to synthesize compounds with carbazole moiety designed to target these kinds of kinases.

**Keywords:** carbazoles; heterocycles; STAT proteins; STAT inhibitors; target STATs; tumoral cells

**Citation:** Caruso, A.; Barbarossa, A.; Carocci, A.; Salzano, G.; Sinicropi, M.S.; Saturnino, C. Carbazole Derivatives as STAT Inhibitors: An Overview. *Appl. Sci.* **2021**, *11*, 6192. https://doi.org/10.3390/app11136192

Academic Editor: Lidia Feliu

Received: 12 May 2021
Accepted: 30 June 2021
Published: 3 July 2021

**Publisher's Note:** MDPI stays neutral with regard to jurisdictional claims in published maps and institutional affiliations.

## 1. Introduction

Over the years, sulphur and nitrogen-containing heterocyclic compounds have attracted particular interest. Many drugs, both of natural and synthetic origin, bear a heterocyclic structure (for example papaverine, theobromine, emetine, theophylline, atropine, codeine, reserpine, morphine, diazepam, chlorpromazine, barbiturates, and antipyrine) [1–6]. The pharmaceutical and pharmacological importance of these compounds resides in their ability to participate in hydrogen bonding with biological substrates (like specific proteins), where the heterocycle core can embody either H-acceptor as in heteroaromatic molecules or H-donor as in saturated *N*-heterocycles. This peculiarity is involved not just in pharmacological properties, but also in the pharmacokinetic behaviour of such drugs [7]. Among the various classes of heterocyclically structured compounds emerges the carbazole class (Figure 1), first isolated by Graebe and Glazer in 1872 from coal tar [8]. In 1965, Chakraborty et al. [9] reported the isolation and the activities of murrayanine (2) from Murraya koenigii Spreng. Since that moment, these compounds represented a great concern due to the attractive structural characteristics and encouraging biological effects displayed by several carbazole alkaloids [10–12].

Indeed, their structural motif is predominant in different artificial materials and naturally occurring alkaloids [12,13]. The majority of carbazole alkaloids were extracted from the taxonomically related higher plants of the genus *Murraya*, *Glycosmis*, and *Clausena* from the *Rutaceae* family. Moreover, these alkaloids were isolated from fungi and algae belonging to *Streptomyces*, *Aspergillus*, and *Actinomadura* species and the ascidian *Didemnum granulatum* [14]. A number of carbazoles derived from plants possess antitumor, psychotropic, anti-inflammatory, antihistaminic, antibiotic, and antioxidant activities [15]. However, carbazole application is not limited to the biological and pharmaceutical field. In fact, due to their properties, they are also employed in the materials science field, as optoelectronic materials, conducting polymers, and synthetic dyes [16,17]. For instance, polyvinylcarbazoles (PVK) [18] have been thoroughly investigated for their employments in photorefractive materials and xerography. In fact, some poly (2,7-carbazole) derivatives (Figure 2) have been used in polymer solar cells [19]. The broad spectrum of their usage includes organic light-emitting diodes as green, red, and white emitters and some substituents at C-2, -3, -6, -7, and -9 positions can be responsible for these molecular and optical properties [20].

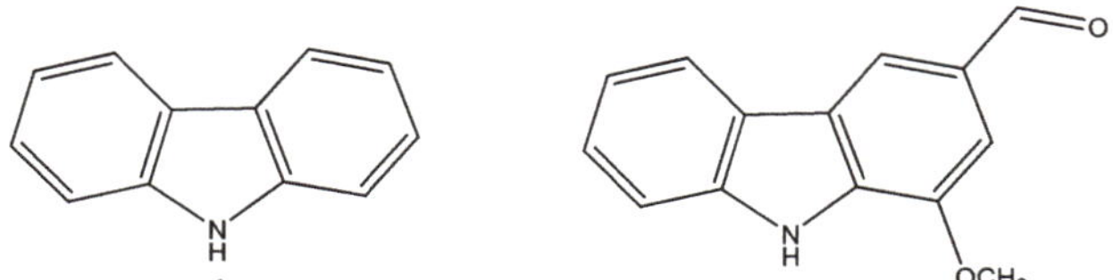

**Figure 1.** Structure of carbazole (**1**) and murrayanine (**2**).

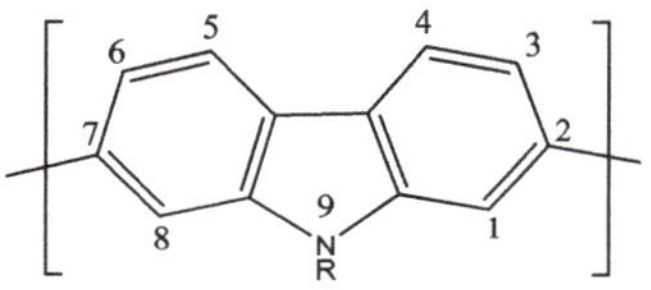

**Figure 2.** Poly (2,7-carbazole).

Many natural carbazoles and their synthetic derivatives contain different substituents on the carbazole ring such as carbazomycin B (**4**) and carbazomadurin A (**5**), while others present a quadricyclic or pentacyclic structure, such as ellipticine (**6**) and staurosporine (**7**) respectively, or can appear as dimers, for example clausenamine A (**8**) (Figure 3) [21].

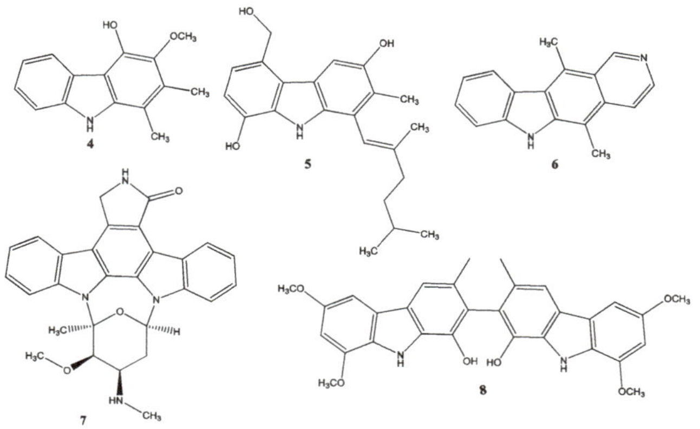

**Figure 3.** Structures of: carbazomycin B (**4**), carbazomadurin A (**5**), ellipticine (**6**), staurosporine (**7**), clausenamine A (**8**).

Due to the growing interest in carbazole bioactivities, different synthetic strategies have been set up and reported in the literature [14,22,23]. Knolker et al. [14] thoroughly reviewed the preparation methods of these carbazole alkaloids. Usually, the synthetic routes to carbazoles include nitrene insertion, Fischer indolization, Pummerer cyclization, Diels-Alder reaction, dehydrogenative cyclization of diarylamines, etc. However, more recently, researchers explored the possibility of using transition metal-mediated C-C and C-N bond formation, cyclotrimerization, benzannulation, Suzuki–Miyaura coupling, and ring-closing metathesis [24–26].

Several studies aimed to compare the structural characteristics of natural carbazole alkaloids. They underlined that 3-methylcarbazole (**9**, Figure 4) could be the essential intermediate in their biosynthesis in higher plants. By contrast, 2-methylcarbazole (Figure 4) could be the communal biogenetic precursor to the tricyclic carbazoles isolated from lower plants. Nevertheless, the natural precursor of the carbazole nucleus has not yet been identified.

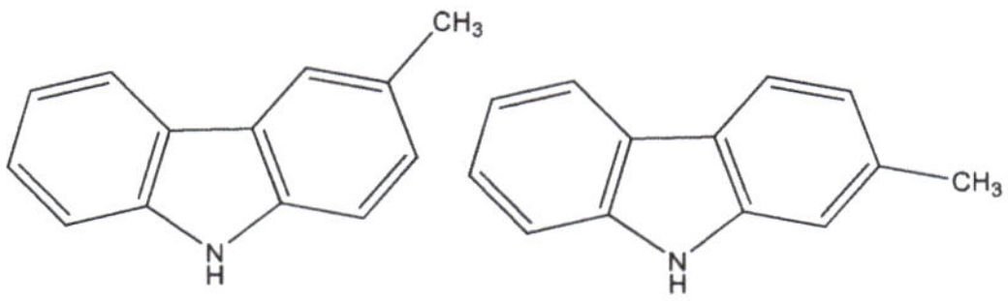

**Figure 4.** Structures of 3-methylcarbazole (**9**) and 2-methylcarbazole (**10**).

Concerning their biological activity, many research groups assessed the antiproliferative action of carbazoles against different types of tumor cells such as breast, cervical, ovarian, hepatic, oral cavity, and small-cell lung cancer [27–30]. One of the principal mechanisms possibly involved in this effect is the ability of carbazoles to target the signal transducers and activators of transcription (STATs) family of signaling pathways (especially STAT3). Indeed, this class of proteins is fundamental for the growth and survival of different human tumor cells [31,32]. In this context, the review aims to examine the studies carried out over the years to develop carbazolic compounds able to act on this target.

## 2. STAT Proteins: Novel Molecular Targets for Cancer Drug Discovery

Signal transducers and activators of transcription (STATs) constitute a group of cytoplasmic proteins that function as signal messengers and transcription factors engaged in critical cellular events with the use of multiple cytokines and growth factors. Since tyrosine has been phosphorylated, two STAT monomers develop dimers via mutual phosphotyrosine-SH2 interactions, migrate to the nucleus, and bind to STAT-specific DNA-response elements of target genes that cause gene transcription.

Concerning the biological functions of STATs, they contribute to cell differentiation, proliferation, development, apoptosis, and inflammation [33,34]. Indeed, STATs are typical transcription factors. They directly engage DNA regulatory elements (DREs) and thereby control the transcription of associated genes, linked to specific functions [35]. Up to now, seven STAT family members have been detected in mammals, designated as STAT1, STAT2, STAT3, STAT4, STAT5a, STAT5b, and STAT6.

The existing connection between the immune system cells and the JAK-STAT pathway has long been studied. Concerning the gain- or loss-of-functions mutation in genes encoding for JAK-STAs, immunodeficiency and susceptibility to infections are the most widespread phenotypes, demonstrating the crucial role of Jak–STAT signaling in the development and function of immune cells. STAT1 signaling, for example, suppresses type 17 immunity, which makes mice more susceptible both to bacterial and viral infections [36]. Indeed, it is involved in poxviruses, which encode interferon-receptor homologs; vaccinia viruses, which encode phosphatases; and Epstein–Barr viruses, encoding a protein that subverts multiple components of the interferon–STAT1 [35]. Additionally, STAT1 exerts a multidirectional antitumor effect. It blocks the cell cycle progression and angiogenesis and induces proapoptotic genes. Moreover, STAT1 signaling governs T helper type 1 (TH1) cell-specific cytokine production that changes both immune functions and inflammatory responses by modifying the ratio between TH1 and TH2 cells. Furthermore, STAT1 has tumor-suppressing properties like TP53 and could be an antagonist for STAT3 and STAT5 activation. Besides, many studies corroborate the relationship between STAT1 downstream of interferons and NF-κB downstream of Toll-like receptors or the cytokine TNF, the two pathways can promote or 'prime' each other [35,37]. STAT2 signaling, instead, is important for its antiviral effects [38]. Further evidence indicates that altered STAT2 signaling may play a role in carcinogenesis by upregulating interleukin-6 (IL-6) production, which activates STAT3 [38].

Previous works demonstrated that in vitro cultured STAT3-deficient T cells did not react to IL-6 stimulation and could not be saved by IL-6 from apoptosis, demonstrating that STAT3's role is crucial for IL-6-mediated anti-apoptotic responses. Further, STAT3 altered functions in keratinocyte physiology could induce skin carcinogenesis. Moreover, constitutively active STAT3 has been identified in several malignancies, among which are breast, melanoma, prostate, head, and neck squamous cell carcinoma (HNSCC), multiple myeloma, pancreatic, ovarian, and brain tumors [39–43].

Aberrant STAT3 signaling stimulates tumorigenesis in part by modifying the expression of genes that rule the tumor growth processes. Examples are offered by genes encoding for p21WAF1/CIP2, cyclin D1, MYC, BCL-XL, BCL-2, vascular endothelial growth factor (VEGF), matrix metalloproteinase 1 (MMP1), MMP7 and MMP9, and survival [44].

A recent study demonstrated that unphosphorylated STAT3 possesses the function of promoting heterochromatin formation in lung cancer cells, suppressing cell proliferation in vitro, and suppressing tumor growth in mouse xenografts [45]. Furthermore, transforming growth factor-α (TGF-α)-mediated epidermal growth factor receptor (EGFR) signaling plays a vital role in the activation of STAT3 in some head and neck cancer cell lines [46]. In addition, STAT3 mutations were identified in multiple peripheral T-cell lymphomas (PTCL), along with high pY-STAT3 expression, particularly in angioimmunoblastic T-cell lymphoma (AITL) and anaplastic large cell lymphoma (ALCL) patient samples, two different PTCL cancer types [47]. By modulating the genetic and pharmacological profile of persistently active STAT3, it is possible to hamper the tumor progression *in vivo*. STAT4 is also essential for IL-12 function, which governs the differentiation of TH1 cells and their inflammatory responses [48]. Therefore, STAT4 signaling is related to autoimmune diseases [49].

STAT5 is present in two isoforms, STAT5a and STAT5b [50]. Normal STAT5 signaling is crucial in mammary gland development, milk production, and hematopoiesis [51]. Constitutive STAT5 triggering is also involved in the development of HNSCC, chronic myelogenous leukemia (CML) [52], and breast, prostate, and uterine cancers [53]. In particular, the effects could be related to the aberrant STAT5 activation by BCR–ABL in CML43 protein. Indeed, it is extensively accepted that STAT5 and STAT3 have in common equal functions in promoting cancer, inducing the expression of pro-proliferative and anti-apoptotic genes [54,55]. Mice lacking both STAT5A and STAT5B showed a large variety of immunological defects and died soon after birth. In addition, the lineage-restricted deletion of STAT5 has been key in defining its in vivo functions, including the role in CD4+ T cell differentiation, NK-cell-mediated immunosurveillance, CD8+ T cell memory187, DC-lineage specification and DC-driven type 2 inflammation. [35] In addition, STAT5A and B are implicated in chronic myeloid leukemia. In fact, the transfection of K-562 leukemia cells either with anti-STAT5A or anti-STAT5B, led to a greatly increased apoptosis rate [56].

IL-4 and IL-13 trigger STAT6 signaling that sustains immune function and regulates the equilibrium between inflammatory and allergic immune responses [57,58]. STAT6 signaling also supports luminal mammary epithelium progression and is involved in the pathology of lung and airway diseases, [59–61]. A recent study assessed that E2F1 is essential for maintaining the level of signal transducer and activator of STAT6 in HCT116 colorectal cancer cells. Mechanistically, E2F1 induced specificity protein 3 (SP3) directly binds to the promoter of the STAT6 gene and activates its transcription in CRC cells. As a result, it was demonstrated that the E2F1/SP3/STAT6 axis is required for the IL-4-induced epithelial-mesenchymal transition of colorectal cancer cells [62].

Linked to STAT proteins are Janus kinases (JAK1, JAK2, JAK3, and TYK2) enzymes, structurally related to each other, with tyrosine kinase activity assisting in the passage of signals from the cell surface to the inside. The JAKs pathways are implicated in the onset of many inflammatory and autoimmune pathologies, among which are rheumatoid arthritis (RA), psoriasis, and inflammatory bowel disease. The main reason is due to the use of the JAK-STAT3 pathway by a substantial amount of cytokines and hormones for intracellular signaling. Indeed, in psoriasis, diverse cytokines implicate the JAK-STAT pathway, such as type I and II IFN, IL-12, IL-22, IL-23, and IL-13. In particular, JAK activation starts, at first, upon ligand-mediated receptor multimerization because two JAKs are brought into proximity, allowing trans-phosphorylation. The subsequent step includes the phosphorylation of a conserved tyrosine residue near the C-terminus of STATs by activated JAKs [63]. Concerning the JAKs family, JAK1, JAK2, and TYK2 expression can be detected throughout the whole body. On the contrary, JAK3 has been specifically identified in the hematopoietic cells. JAK1 displays its function by conveying the signals of numerous proinflammatory cytokines and cooperates with JAK3 in lymphopoiesis by binding to heterodimeric interleukin (IL) receptors. JAK2 is one of the major mediators in the signal transmission of hematopoietic factors. Moreover, recently, a gain-of-function point mutation in the pseudokinase domain (JH2) of JAK2 has been ascertained in myeloproliferative neoplasm (MPN) patients [64]. JAK3 and TYK2, instead, possess immunomodulatory functions [65].

### 3. STATs Inhibitors

#### *3.1. Peptides and Small Molecules as STATs Inhibitors*

Activation of STAT1, STAT3, and STAT5 is highly frequent in almost all tumors studied, with a higher incidence of abnormal STAT3 activation [66]. Starting from the breakthrough of the first peptide inhibitor of a STAT protein, diverse strategies to counteract against STAT signaling have been pursued. An example is provided by small-molecule dimerization disruptors (SMDDs) or phospho-peptidomimetic inhibitors (PPMIs) targeting the phospho-Tyr-SH2 domain interaction at the interface of dimers of STAT proteins. The main consequence consists of the disruption of STAT–STAT dimers. STAT–SMDD or STAT–PPMI forms heterocomplexes, suppressing STAT signaling and function. On the other hand, the bond between growth factors and cytokines and their receptors on the cell surface activates STAT tyrosine phosphorylation. Tyrosine kinases responsible for STAT phosphorylation are the targets of small-molecule tyrosine kinase inhibitors. These modulators can prevent the induction of STAT phosphorylation and signaling. They act by impeding STAT dimerization using peptides or peptidomimetics identified through different processes as structure-based design, small molecules identified by molecular modeling, virtual or library screening, or natural products [66].

Other methods, comprise oligodeoxynucleotide (ODN) decoys (peculiar STAT DNA-binding domain inhibitors) and antisense oligonucleotides (ASOs) that interfere with STAT mRNA. Studies on the molecular basis of oncogenesis, regarding oncoproteins like v-Src, report alterations in intracellular signaling proteins involved in several malignancies. The discovery that STAT3 is constitutively activated in v-Src transformation indicated the possible central role of STATs in oncogenesis [67,68]. Additionally, other transforming tyrosine kinases such as v-Eyk [69], v-Ros [70], v-Fps [71], Etk/ BMX [72], and Lck [73], activate STAT3 in the oncogenic process. Constitutive STAT3 activation is also connected with the transformation caused by tumor viruses, among which are HTLV-1 [74], polyomavirus middle T antigen [75], EBV [76], and herpesvirus saimiri [73], that directly or indirectly activate JAKs or Src family tyrosine kinases.

For the reasons explained above, different therapeutic agents, targeting aberrantly active STAT3, could affects several human cancers making STAT3 the target in many drug discovery research efforts. STAT3 drug discovery research concentrated on targeting the pTyr-SH2 domain interaction [77,78] due to its relevance in promoting STAT3 dimerization and function. A semirational, structure-based design study determined the first SH2 domain-binding peptides and peptidomimetics that break up the STAT3 pTyr-SH2 domain interactions and STAT3–STAT3 dimerization [78,79]. The native parent pTyr peptide, PY*LKTK (where Y* stands for pTyr) and its modified forms prevented the DNA-binding and transcriptional activities of STAT3 at high doses [79]. Peptidomimetic and non-peptide analogues, such as ISS-610 and S3I-M2001, counteracted STAT3 activity in vitro and accommodated aberrantly active STAT3 [78,79]. In particular, S31-M2001 constrained the progression of human breast tumor xenografts, while ISS-610 blocked cell growth and triggered apoptosis in vitro [66].

Moreover, phosphopeptide binding sequences with the primary structure pTyr-Xxx-Xxx-Gln (where Xxx represents any amino acid) impeded STAT3 activation. They derived from leukemia inhibitory factor (LIF), IL-10 receptor, epidermal growth factor receptor (EGFR), granulocyte colony-stimulating factor (GCSF) receptor, or glycoprotein 130 (gp130) [80,81]. In this research, the peptidomimetic Ac-pTyr-Leu-Pro-Gln-Thr-Val-NH$_2$ inhibited STAT3 activity (IC$_{50}$ values of 150 nμM) [81,82]. Furthermore, a 28-mer native peptide identified as SPI, obtained from the STAT3 SH2 domain, hampered the STAT3 pTyr-SH2 domain interaction and signaling. It suppressed cell viability and caused programmed cell death of human breast, pancreatic, prostate, and non-small cell lung cancer (NSCLC) cells in vitro [83].

In silico studies of chemical libraries detected several small-molecule inhibitors of STAT3 activity. The main mechanism regards the disruption of STAT3–STAT3 dimerization. STA-21 (or NSC628869) was recognized from the screening of the National Cancer Institute (NCI) chemical library as an inhibitor of STAT3 dimerization, DNA-binding activity, and transcriptional function in breast cancer cells at 20 μM [84]. Its structural analogue, LLL-3 (with enhanced membrane permeability), reduced cell viability in vitro and intracranial tumor size in vivo in glioblastoma animal models [85]. Moreover, a catechol (1,2-dihydroxybenzene) compound was detected from Wyeth's proprietary small-molecule collection as a STAT3 SH2 domain inhibitor, active at 106 μM against a multiple myeloma cell line [86]. In addition, S3I-201 (or NSC74859) as a STAT3–STAT3 dimerization disruptor active at 60–110 μM [87] also emerged from the NCI chemical library. S3I-201 hampered STAT3 DNA-binding and transcriptional activities, inhibited cellular growths, and promoted the apoptotic process of tumoral cells harboring constitutively active STAT3, and suppressed the growth of human breast cancer xenografts [87]. Several of its derivatives, such as S3I-201.1066, BP-1-102, and S3I-1757, exhibited a better efficacy, with IC$_{50}$ values of 35 μM (S3I-201.1066), 6.8 μM (BP-1-102), and 13.5 μM (S3I-1757), and suppressed cell growth, malignant transformation, survival, migration and invasiveness in vitro of malignant cells harboring aberrantly active STAT3.

Moreover, Cpd30 (4-(5-((3-ethyl-4-oxo-2-thioxo-1,3-thiazolidin-5-ylidene)methyl)-2-furyl)benzoic acid) (Figure 5) moderately blocked STAT3, prevented STAT3 nuclear translocation upon IL-6 stimulation, and triggered apoptosis in breast cancer cells harboring constitutively active STAT3 [54]. Its analogue, Cpd188 (4-((3-((carboxymethyl)thio)-4-hydroxy-1-naphthyl)amino)sulphonyl)benzoic acid) (Figure 5), combined with docetaxel, lowered tumor growth in chemotherapy-resistant human breast cancer xenograft models.

From another virtual screening of small molecules emerged Stattic, a non-peptide small molecule able to target the STAT3 SH2 domain and inhibit STAT3 signaling at 10 μM [88]. It triggered apoptosis of STAT3-dependent breast cancer and neck squamous cell carcinoma HNSCC cells and arrested growth of orthotopic HNSCC tumor xenografts [88]. STX-0119, STAT3 SH2 domain antagonist generated antitumor effects both in vitro and in vivo in a human lymphoma model, probably by disrupting STAT3–STAT3 dimerization, with a modest activity on STAT3 phosphorylation [89]. Fragments of STX-0119 and stattic were chemically combined to create HJC0123 (Figure 5) [90], which impeded STAT3 phosphorylation and transcriptional activity in breast cancer cells and possessed antiproliferative effects towards breast and pancreatic cancer cells in vitro (IC$_{50}$ values of 0.1–1.25 μM).

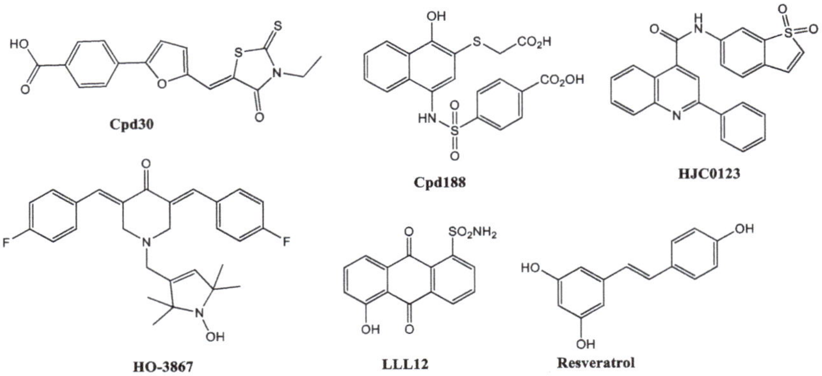

**Figure 5.** Some inhibitors of STAT3 activity with no carbazole structure.

The oral administration of HJC0123 blocked the growth of human breast cancer xenografts [90]. Furthermore, OBP-31121 inhibited STAT1, STAT3, and STAT5 phosphorylation. OBP-31121 reduced cell proliferation and triggered apoptosis in vitro whereas it prevented tumor growth in vivo in gastric cancer models, and it further sensitized gastric cancer cells to cisplatin and 5-fluorouracil [91].

### 3.2. Natural Compounds as STATs Inhibitors

Several studies also identified many lead candidates such as STAT3 inhibitors from natural products. However, the mechanism of action is not completely defined. Curcumin, a phenolic compound derived from the perennial herb *Curcuma longa,* hampers JAK–STAT signaling at 15 μM, promotes cell cycle arrest, and impedes cell invasion in vitro in a small cell lung cancer model [92]. The treatment of mice bearing gastric cancer xenografts with curcumin suppressed IL-6 production by IL-1β-stimulated myeloid-derived suppressor cells, associated with decreased activation of STAT3 and nuclear factor-κB (NF-κB).

Curcumin analogues with ameliorated bioavailability and stability, such as FLLL32, had improved efficacy ($IC_{50}$ values of 0.75–1.45 μM) in inhibiting both pSTAT3 and total STAT3. Furthermore, they caused STAT3 ubiquitylation and possible proteasomal degradation in canine and human osteosarcoma cells in vitro [93]. HO-3867 (Figure 5), curcumin analog, downregulated likewise STAT3 signaling in cisplatin-resistant human ovarian cancer cells, thus intensifying sensitivity to cisplatin [94]. HO-3867 also promoted apoptosis in BRCA1-mutated human ovarian cancer cells harboring aberrantly active STAT3 [95].

Other findings estimated that LLL12 (Figure 5), another small-molecule inhibitor of STAT3 signaling based on curcumin [96], hampers STAT3 activation by impeding its recruitment to the receptor and avoiding phosphorylation by tyrosine kinases, and by preventing the dimerization [97]. LLL12 decreased cell viability, promoted the apoptotic process, and suppressed colony formation and migration in vitro in glioblastoma, osteosarcoma, and breast cancer cells. It also interfered with angiogenesis, tumor vasculature development, and tumor growth in vivo in osteosarcoma xenograft models [96,97]. Resveratrol (3,5,4′-trihydroxystilbene) (Figure 5), a polyphenolic compound mainly found in red grapes, inhibits STAT3 signaling at high micromolar concentrations, thus affecting tumoral cell growth. Resveratrol treatment also prevents constitutive and IL-6-induced STAT3 activation in multiple myeloma, leukemia, and other tumor cell types, reduces the expression of BCL-2 and other anti-apoptotic proteins, in vitro [98]. The resveratrol derivative LYR71 arrested cell growth ($IC_{50}$ value of 20 μM) and decreased STAT3-mediated MMP9 expression [99]. Moreover, the inhibition of the JAK–STAT3 pathway by resveratrol or its analogue, piceatannol (3,3′,4,4′-transtrihydroxystilbene), reduced BCL-XL and BCL-2 expression and sensitized lung carcinoma, multiple myeloma, prostate pancreatic cancer, and the glioblastoma multiforme patient-derived CD133-positive cells to radiation or chemotherapy in vitro.

*3.3. Carbazoles as STAT Inhibitors*

Psoriasis, a common inflammatory skin disease, is treated, according to its severity, with a large panel of therapies associated with side effects. Coal tar represented an ancient treatment of psoriasis. To assess the potential mechanism of action, Arbiser et al. in 2006 [31] fractionated coal tar through chromatography determining carbazole as the active compound. STAT3 inhibition resulted in one of the main mechanisms implied in its antiangiogenic activity. Indeed, STAT3 is involved as a signaling pathway in psoriasis. Moreover, carbamazepine (Figure 6) (an anticonvulsant structurally related to carbazole) exhibited great potentiality in patients with generalized psoriasis. For this reason, this research group focused their attention on carbazole **1** included in coal tar.

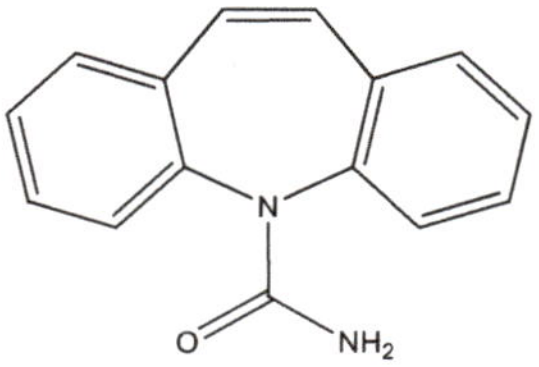

**Carbamazepine**

**Figure 6.** Structure of carbamazepine, an anticonvulsant structurally related to carbazole.

Furthermore, carbazole inhibited IL-15, whose level is increased in psoriasis, and decreased the effects of inducible nitric oxide synthase (iNOS). STAT3, IL-15, and iNOS need rac GTPase activation to exert their activities. Thus, the ability of carbazole to inhibit rac activation has been demonstrated, stating that this could be the origin of its ability to reduce inflammation and counteract angiogenesis [31]. Starting from the assumption that STAT3 is one of the main mediators implied in the immune and angiogenic features of psoriasis, Arbiser et al. [31] went into the possible actions of carbazole and 2-hydroxycarbazole (a carbazole metabolic by-product) toward the STAT3-mediated transcription. Their data, obtained on RAW264.7 macrophages, indicated that both carbazole and 2-hydroxycarbazole (**11**, Figure 7) could downregulate STAT3-mediated transcription.

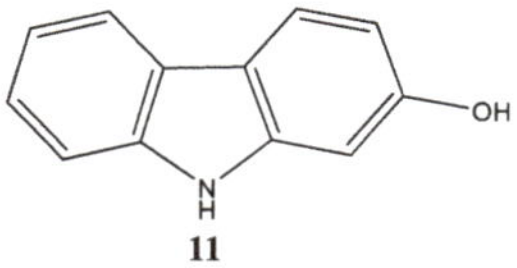

**11**

**Figure 7.** Structure of 2-hydroxycarbazole (**11**).

They assessed the specificity against STAT3, since carbazole possessed no activity on a constitutive promoter. Activity on STAT3 phosphorylation was excluded as a potential carbazole mechanism of action. Indeed, no effect on STAT3 phosphorylation was detected, implicating the existence of an alternative mechanism. A fundamental prerequisite to activate STAT3 is rac activation. In fact, small GTPase rac influences STAT3 transcription, IL-15 production, and iNOS activity. Obtained results outlined that carbazole markedly hampered rac activation by VEGF in HUVEC cells.

In 2013, Sarkar et al. [100] analyzed the molecular mechanism held responsible for the antiproliferative activity of mahanine (**12**, Figure 8) toward several pancreatic cell lines. In fact, natural compounds have been demonstrated to hold a great deal of promise as antineoplastic agents. In particular, the purified carbazole alkaloid mahanine, isolated from the edible plants *Murraya koenigii* and *Micromelum minutum*, displayed antimutagenic, antimicrobial, and cytotoxic activities [101].

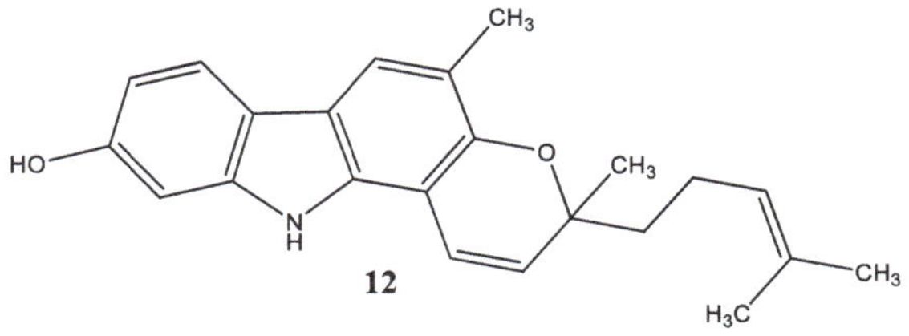

Figure 8. Structure of mahanine (**12**).

Furthermore, mahanine triggered the apoptotic process in histiocytic lymphoma, promylocytic leukemia, and prostate cancer cells [102–104]. At first, it was confirmed that mahanine induced apoptosis in an in vitro model of the human pancreatic ductal adenocarcinoma cell line, MIAPaCa-2 (IC$_{50}$ amounting to 13.9 μM). Subsequently, the possible effects of mahanine on the molecular chaperone Hsp90 have been studied, starting from the assumption that it presents high expression levels in several tumor cells and regulates different client proteins, including STAT3. The results underlined that mahanine triggers the accumulation of ROS in both a time- and concentration-dependent manner in MIAPaCa-2 cells. The main consequence of this activity regards the oxidative insult of Hsp90, mainly involved in maintaining folded proteins in their proper conformation. Immunostaining studies demonstrated the depletion of the two Hsp90 client proteins Akt and STAT3 in orthograft pancreatic adenocarcinoma mouse model treated with mahanine. Furthermore, mahanine lowered, in a concentration-dependent way, other client proteins such as B-Raf, A-Raf, mutated p53, GSK3b, and PKCb at 24 h in MIAPaCa-2 and AsPC-1 up to a dose of 30 μM. Besides, the authors proved to decrease mahanine concentration (10–20 μM), demonstrating its efficacy in MIAPaCa-2 cells after 24 h of treatment. In fact, mahanine diminished the protein level of Akt and CDK-4 at 15 μM, while B-Raf, STAT3 and Bcl-XL were conspicuously reduced at 17.5 μM. As previously mentioned, the molecular chaperone Hsp90 has a crucial function in the survival of cancerous cells in which it is extensively expressed. As a result, this makes it a promising target for chemotherapeutic agents. Indeed, it mediates the folding, assembly, and maturation of many client proteins, among which are HER2, EGFR, PI3K, Akt, B-Raf, STAT3, GSK3b, Cdk4, mutated p53, and steroid receptors that participate in malignant cancer development [105].

In silico studies demonstrated the binding capability of mahanine to Hsp90 via noncovalent weak interactions, in a polar pocket, apart from the ATP-binding cavity. Mahanine constituted two hydrogen bonds with Hsp90, the first with the side chain oxygen of Glu47 through its NH group and the second with the side chain oxygen of Asn51 through its OH group. Moreover, van der Waals interactions with Arg46, Ile43, Gly132, Gln133, Met130, Ser129, Phe138, Ile131, and Gly137 were established. The most advantageous configuration of the ligand with the protein had a binding energy of $-7.6$ kcal mol$^{-1}$, with a micromolar binding affinity ($K_D = 3.16$ μM) [100].

In 2013, Saturnino et al. [32], in their preliminary study, synthesized a series of *N*-alkylcarbazole derivatives to evaluate their potential STAT3 inhibitory activity. Their studies, indeed, highlighted that a crucial role in modulating the lipophilic properties of the carbazole was played by the substituent in N9. They analyzed a small series of carbazoles *N*-alkylated with C5, C6, and C7 alkyl chains [106]. Additionally, the alkyl chains were functionalized with dimethyl 5-hydroxyisophthalate (for derivatives **13**) or methyl salicylate (for derivatives **14**) as substituents (Figure 9). Their findings highlighted the ability of compounds **13a-c**, among others, to inhibit STAT3 phosphorilation and its nuclear translocation in acute monocytic leukemia at 50 μM with different potencies, amounting to 50%, 90%, and 95%, respectively. A crucial element for this effect seems to be the length of an alkyl linker inserted in these compounds. They obtained these outcomes by performing

EMSA and Western blot analysis in THP-1 cells treated with IL-6 (20 µg/mL) for 15 min. Indeed, IL-6 augmented the STAT3 DNA binding activity as indicated by EMSA/supershift experiments with anti-STAT3 antibody. Furthermore, obtained data suggested a time- and dose-dependent effectiveness for **13a-c**. These compounds, however, could not block IFNy-induced STAT1 nor TNF-α LPS-induced NF-kB activation. Phosphorylation on specific tyrosine residues and consecutive translocation into the nucleus are crucial to activate STATs. Their studies assessed that compounds **13a-c** lowered, at different levels, IL-6-induced tyrosine705 phosphorylation of cytosolic STAT3 without impacting the whole STAT3 protein. Concerning the structure-activity relationship, the insertion of an *N*-alkyl chain added an interaction point to the target protein.

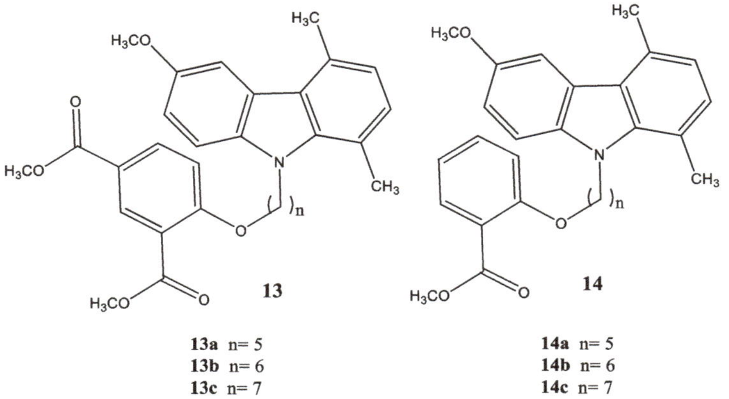

**13a** n= 5
**13b** n= 6
**13c** n= 7

**14a** n= 5
**14b** n= 6
**14c** n= 7

**Figure 9.** Structure of: hydroxyisophthalate derivatives (**13**) and methyl salicylate derivatives (**14**).

Slight changes of its terminal phenyl ring significantly modified the activity profile (compare **13a-c** with **14a-c**). Another important feature to improve the activity is the length of the alkyl linker as observed by comparing **13a** with **13b** and **13c**) [15,32].

Although prior research substantiated that [31,32] some carbazoles affected the STAT3-mediated transcription and/or STAT3-DNA binding and phospho-STAT3 at high doses (30–50 µM), little evidence regards the molecular mechanisms involved in this process. Hou et al. [107] have been previously fascinated by the possibility of targeting STAT3 pathways. They, indeed, tried to identify therapeutic agents against cancer. They focused on the development of fluorescent small molecules. The silver lining of these compounds was the capability to be observed directly as it interacts with receptor-positive cell lines [108–110]. However, a preceding survey assessed that the antiproliferative activity of the free carbazole on the cancer cell lines is lower than the corresponding scaffolds with the dansyl group. In view of the above, this research group synthesized a small series of carbazole derived compounds with fluorophore (Figure 10). Among them, compound **15** was the most effective compound.

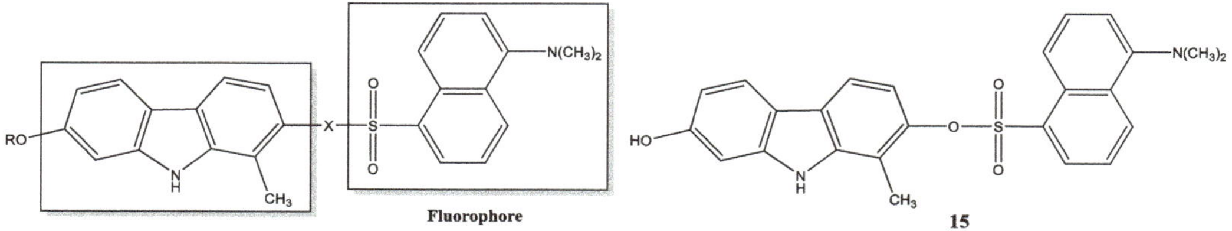


**Figure 10.** General structure of carbazoles fluorophore and structure of 7-hydroxy-1-methyl-9H-carbazol-2-yl 5-(dimethylamino)-naphthalene-1-sulfonate (**15**).

It inhibited the STAT3-mediated transcription and IL-6-induced phosphorylation of STAT3 in triple negative breast cancer (TNBC) cells. Indeed, they examined the antiproliferative effects of **15** on a panel of breast cancer cells, finding the best results for the invasive ductal carcinoma cells SUM1315MO2. Moreover, **15** suppressed cell proliferation in A431 (a squamous carcinoma cell line), A549 (a lung cancer cell line) and PC-3 (a prostate cancer cell line) with GI50 values 0.16 µM (for A431), 2.5 µM (for A549), and 3 and 7.9 µM (for PC-3), respectively.

According to the inhibition of STAT-3, the cell treatment with compound **15** provoked a reduction of cyclin D1 levels, transcriptional target of STAT3. Besides, in vitro and in vivo studies demonstrated that the inhibitory effects of **15** on phospho-STAT3 were via the up-regulation of cytoplasmic protein−tyrosine phosphatase PTPN6. This was measured on the two lines of invasive ductal carcinoma HS578T, and SUM149PT. Further analysis assessed a cytoplasmic fluorescence of compound **12** in MDA-MB-231 cells, evidencing that the compound enters human breast cancer cells. Compound **15**, in fact, triggered apoptosis in breast cancer cell lines in vitro and it was efficient at inhibiting the in vivo growth of human TNBC xenograft tumors (SUM149) with a 57% reduction of tumor volume on the fifth day without showing any toxicity. Furthermore, compound **12** also suppressed the growth of human lung tumor xenografts (A549) harboring aberrantly active STAT3. The maximum inhibitory effect was achieved on the ninth day (inhibition rate 73.76%) [107].

Subsequently, in 2017, the same research group focused their attention on this compound deepening the knowledge of the molecular mechanism [111]. Indeed, as just reported, in their preceding work they discovered that **15** was able to inhibit phospho-STAT3 (Y705) by induction of PTPN6/SHP-1 expression. However, previous studies underlined that PTPN6/SHP-1 is epigenetically repressed by STAT3- DNMT1 [112] and acetylation of STAT3 (K685) is critical to bind DNA (cytosine-5)-methyltransferase 1 (DNMT1). They found that **15** not only decreased the phospho-STAT3 (Y705) but also the acetyl-STAT3 (K685) levels in HS578T cells, both in a dose-dependent manner.

By using Western blot experiments, they highlighted that STAT3 binds to DNMT1 in HS578T cells. However, **15** inhibited the capability of forming this interaction at 0.3 µM. The disruption of STAT3-DNMT1 interaction by **15**, instead, happened at 0.1 µM. This effect was achieved also in SUM1315MO2 cells. In contrast, the reference molecules used, the JAK2 inhibitor (AZD1480) or JAK1/2 inhibitor (CP690550) did not impact STAT3-DNMT1 interaction. Moreover, they established that the disruption of STAT3- DNMT1 interaction by **15** is dependent on the deacetylation of STAT3 at K685. Furthermore, starting from the assumption that the STAT3-DNMT1 interaction is important for DNA methylation in the promoter region of TS genes, they identified four TS genes that were demethylated by **15** in HS578T cells, among which the retinoic acid receptor beta (RARB), neurogenin 1 (NEUROG1), PDZ and LIM domain 4 (PDLIM4), and Von Hippel–Lindau tumor suppressor (VHL). Then, they evaluated the methylation status of these promoters in two TNBC cells. It has been discovered that the promoter regions of VHL and PDLIM4 genes were highly methylated and **15** near commonly demethylated these promoters. Demethylation of VHL and PDLIM4 gene promoters by **15** resulted in the reactivation of mRNA expression of these genes in both HS578T and MDA-MB-231 in a dose-dependent manner. At the end, they carried out in vivo studies, highlighting that **15** significantly reduced tumor growth without inducing loss of body weight [111].

In 2015, Cuenca-López et al. [113] studied the antitumor properties of **16** (Figure 11), a hybrid indolocarbazole analog produced by combinatorial biosynthesis of Rebeccamycin and Staurosporine genes.

The authors outlined the antiproliferative activity of this compound towards HS578T, BT549, MDA-MB-231, and HCC3153 with $IC_{50}$ values in the nanomolar range. Moreover, **16** was able to reduce colony formation of these cells and decrease tumor volume in mice. Besides, their main purpose was to investigate the molecular mechanism of **16** by analyzing its effects on kinases profile in TNBC. Through biochemical experiments, they proved that **16** suppressed downstream components of the PI3K/AKT pathway including

AKT (phosphorylated at T308 and S473) and pS6. In the same way, it suppressed p-Stat3 and p-Stat1 in HS578T and BT549 cells. To deepen their knowledge of the antiproliferative mechanism, they carried out some experiments on cell cycle and apoptosis. Their findings suggested that it caused an accumulation of cells HS578T, BT549, and MDA-MB-231 in G2/M phase at 24h whereas further analysis underlined that the apoptotic mechanism was caspases independent. Therefore, to examine if **16** induced DNA damage, they evaluated the phosphorylated γH2AX levels, a protein necessary for checkpoint-mediated cell cycle arrest and DNA repair following double-stranded DNA breaks. Treatment with **16** in HS578T, BT549, and MDA-MB-231 exhibited an enhancement in the phosphorylated levels of γH2AX at early time points. They also noted that **16** provoked the phosphorylation of p53 and Chk2, validating the induction of DNA damage. In addition, they tried to explore the effects of **16** when associated with chemotherapeutic agents used in the clinical setting for triple negative tumors such as vinorelbine, docetaxel, and carboplatin. Their results displayed a synergistic interaction of **16** with all these compounds, mostly with docetaxel, in HS578T, BT549, and MDA-MB-231 cells and improved effects *in vivo*. In addition, **16** possessed a good pharmacodynamic and pharmacokinetic profile. Indeed, it exhibited a clear decrease of pS6 and pSTAT1 at 30 min in vivo in the extracted tumors. pSTAT3 was also moderately impaired. Induction of pγH2AX was reached at 60 min. Moreover, they evaluated the concentration of the drug in two resected tumors per time point, assessing a time-dependent accumulation, reaching over 1000 ng/g. The obtained results underlined that targeting acetylation of STAT3 (K685) by small molecule inhibitors could be a valid therapeutic option for cancer treatment. Acetylation of STAT3 and DNMT1 present high levels in a wide range of malignancy. Conceivably, in these diseases, the STAT3-DNMT1 complex is engaged in the repression of TS genes by DNA methylation. With this in mind, an in-depth investigation of acetyl-STAT3 and DNMT1 status in human cancers could offer a potential therapeutic opportunity with the aim of targeting the STAT3-DNMT1 interaction [113].

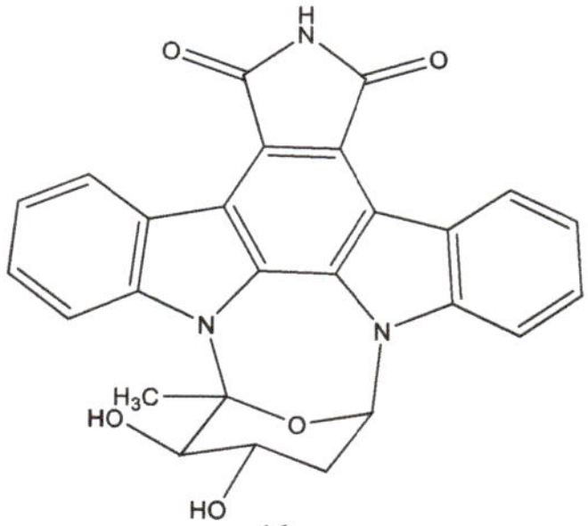

**Figure 11.** Structure of (5*R*,7*S*,9*S*)-7,8-dihydroxy-9-methyl-6,7,8,9-tetrahydro-5*H*,14*H*-17-oxa-4b,9a,15-triaza-5,9-methanodibenzo[b,h]cyclonona[jkl]cyclopenta[e]-as-indacene-14,16(15*H*)-dione (EC-70124, **16**).

Oscar Estupiñan et al. in 2019 [114] evaluated the anticancer effect of **16**, too. They assessed that PI3K/AKT/mTOR, JAK/STAT or SRC were the main activated pathways in cell-of-origin sarcoma models and/or sarcoma primary cell lines. Compound **16** blocked the phosphorylation of these targets and impeded proliferation triggering the DNA damage, cell cycle arrest, and apoptosis. Moreover, **16** in part decreased tumor growth *in vivo*. Furthermore, it decreased the expression and the effect of ABC efflux pumps implicated in drug resistance. Data also suggested that combining compound **16** with doxorubicin resulted in a synergistic cytotoxic effect in vitro and an enhanced antitumor effect in the animal model. They found that **16** is much more successful than midostaurin in preventing the phosphorylation of ERK1/2 at Thr202 and Tyr204, AKT at Ser473 and Thr308, pS6 at Ser235 and Ser236 and 4EBP at Ser65, which are the kinases/kinase substrates most enabled among a panel of relevant signaling molecules in a cell model of MRCLS. Furthermore, **16** prevented the activation of SRC, STAT1, and STAT3, activated in several primary sarcoma cell lines. The multikinase inhibitory effect of **16** in sarcomas is linked to a potent antiproliferative activity if compared with midostaurin. This is due to the induction of DNA damage followed by a more stringent S-phase arrest and apoptosis. In addition, it is known that midostaurin and other indolocarbazoles inhibit Aurora kinase abrogating the mitotic spindle checkpoint and the accumulation of cells with 4 N and 8 N DNA content [114].

In 2015 Botta et al. [115] synthesized a series of 1,4-dimethyl-carbazole derivatives (Figure 12) and tested them for their potential activity on STAT3.

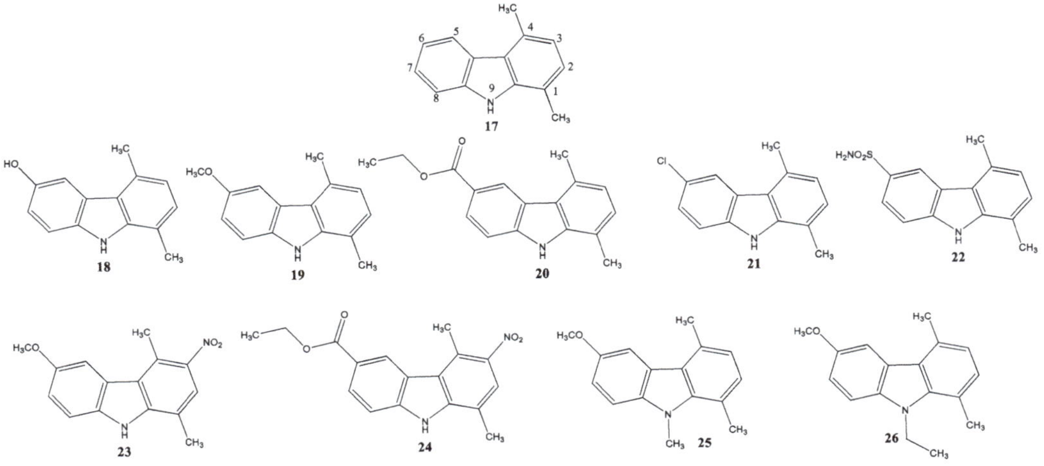

**Figure 12.** 1,4-Dimethyl-carbazole derivatives (**17–26**).

The design of molecules has been assisted by in silico studies with the purpose to identify new ligands of SH2 domain of STAT3 responsible for the protein dimerization, where a monomer identifies the Pro-Tyr/pTyr-Leu-Lys-Thr-Lys sequence of its partner. Indeed, the in silico analysis was performed by selecting the protein region accommodating the Pro-Tyr/pTyr-Leu-Lys-Thr-Lys sequence and especially the interactions involving the Tyr705/pTyr705, essential for the possible ligand binding. Therefore, the possible effects on the cellular viability of human melanoma (A375) and human epithelial cervix adenocarcinoma (HeLa) cell lines, which constitutively express STAT3, have been evaluated. Their findings evidenced that 1,4-dimethyl-carbazole (**17**) (IC$_{50}$ of 80.0 µM) had similar activity to doxorubicin (IC$_{50}$ of 87.0 µM), used as a reference molecule, toward A375 cells.

Compounds **18**, **20**, **24**, and **26** achieved a similar trend of results on both cell lines in comparison with doxorubicin whereas compounds **19** and **21** exhibited IC$_{50}$ values on A375 of 50 and 60 µM, respectively. Moreover, the most active compounds **19–21** displayed a considerable inhibition rate at 72h of STAT-3 expression with respect to the control cells, with a percentage of inhibition amounting to 94.33, 90.66, and 91.0 %, respectively. In silico studies revealed that the position 6 of carbazole ring faces the cavity accommodating the Tyr705/pY705. Specifically, this research group verified that the Tyr705/pY705 is in a small hollow, to interact with the side chains of Lys591, Arg609, Ser611, Glu612 and Ser613. For this reason, they designed compounds **18–22** inserting at C-6 a hydrogen bond acceptor to replicate this interaction. Furthermore, they added hydroxyl, methoxy, ethyl ester groups, and a chlorine and a sulfonamide function. Since position 3 of carbazole is close to the side chain of Arg595, they explored the C-3 position to ameliorate the binding to the protein. Therefore, in light of the simple introduction of a nitro group able to interact with Arg595, compounds **23** and **24** have been designed. Furthermore, compounds **25** and **26** attempted to raise the contacts with STAT3 through van der Waals interactions by introducing alkyl groups on the nitrogen to the detriment of the hydrogen bond. At the end, the designed compounds **18–26** were docked into the binding cavity of STAT3. The docked poses of **18–26** accommodated well into the protein binding cavity and seem to respect the predicted interactions [115].

Since JAK and STAT proteins (as previously stated) are connected to a pathway that involves many cytokines and hormones for intracellular signaling with potential involvement in cancer, Zimmermann et al. [116] designed and synthesized a series of 9*H*-carbazole-1-carboxamides (Figure 13) in order to evaluate their potentiality as selective ATP-competitive inhibitors of Janus kinase 2 (JAK2).

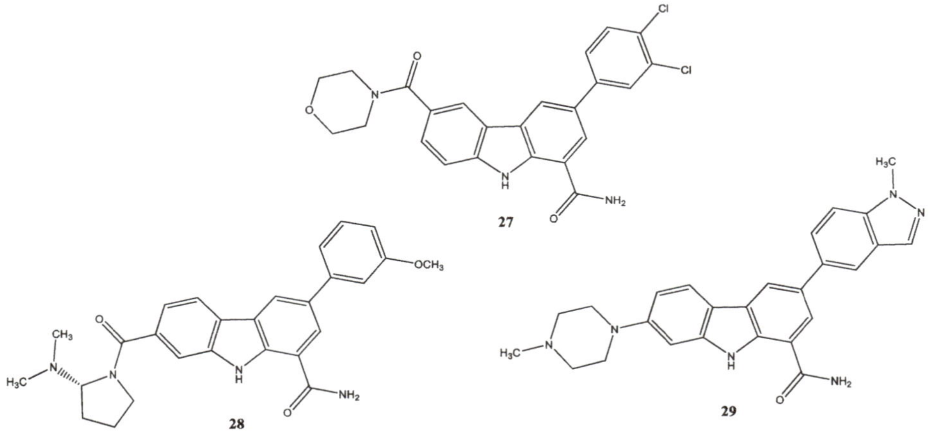

**Figure 13.** Structure of 9*H*-carbazole-1-carboxamides (**27–29**).

By means of in silico studies, they tried to optimize their lead compound **27**, showing a good JAK2 inhibitory potency but poor selectivity toward JAK family. The (*S*)-dimethylamino-pyrrolidine amide **28** exhibited an enhanced selectivity against other members of the JAK kinase family. Moreover, **28** was tested in vitro for its antiproliferative effect on SET-2 cells (adult acute megakaryoblastic leukemia) reaching an $IC_{50}$ value of 80 nM. Further studies were aimed to assess its in vitro ADME properties. Data suggested that **28** possessed a greater solubility than compound **27** (>1.9 mg/mL at pH 1.0; 0.41 mg/mL at pH 6.5; compared to <0.001 mg/mL across pH range for 1). Another compound with a significant selectivity on JAKs was **29**, particularly active against JAK2 ($IC_{50}$ amounting to 5.5 nM), which also displayed an antiproliferative activity against SET-2 cells with an $IC_{50}$ value amounting to 130 nM. Concerning the metabolic stability of **29**, it is similar to **28**, except its solubility at acidic pH was enhanced when compared to the initial screening (0.044 mg/mL at pH 1.0) [108].

Based on the knowledge that phosphorylation of STAT3 at Y705 induces the dimerization and translocation into the nucleus to relay the oncogenic signals by expressing the genes implicated in proliferation, antiapoptosis, angiogenesis and tumor evasion in 2016, Baburajeev et al. [117] reported the synthesis and evaluation of substituted carbazole derivatives (Figure 14) using nano-cuprous oxide as a catalyst via intramolecular C–N bond forming reactions. Among them, (3-acetyl-6-chloro-9*H*-carbazol-9-yl)methyl)-[1,10-biphenyl]-2-carbonitrile (**30**) exhibited the greater antiproliferative activity towards two lung cancer cell lines: A549 and LLC, establishing itself as a lead compound. The calculated $IC_{50}$ values amounted to 13.6 and 16.4 µM, respectively for the two cell lines.

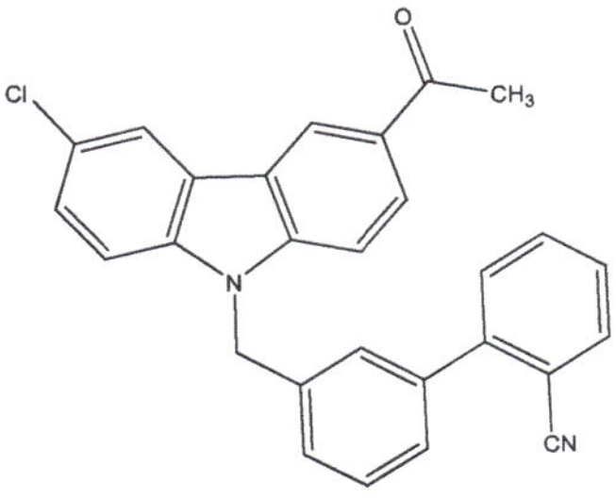

**Figure 14.** Carbazol-carbonitrile derivatives (**30–32**).

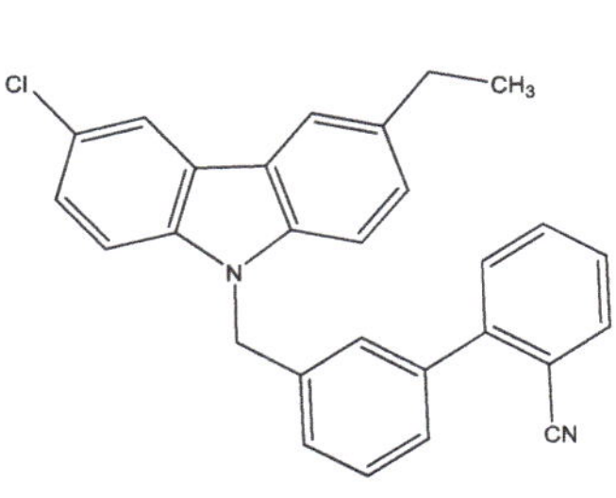

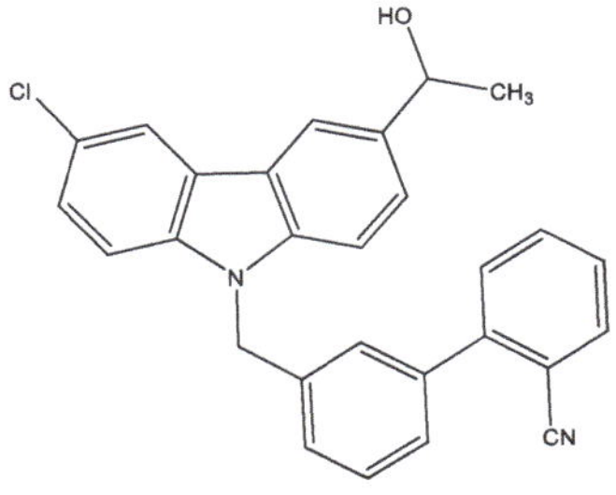

Paclitaxel, used as a reference molecule, instead achieved $IC_{50}$ of 0.0044 µM against A549 cells. Furthermore, they performed their studies on A549, HCC-2279 and H1975 cells, treating them with 10 µM of ACB for 6 h to examine the effect on the persistent activation of STAT3 by Western blot analysis using antibodies that recognize phosphorylation of STAT3 at Tyr-705. They found that **30** inhibited STAT-3 phosphorylation in A549, HCC-2279 and H1975 cells at a concentration of 10 µM. However, **30** did not affect total STAT3 and -actin levels, confirming the inhibitory effects of carbazoles on STAT3 phosphorylation. Moreover, they analyzed the levels of phospho-STAT3 and lysine demethylase (LSD1) in the nuclear extract of HCC-2279 cells, at different concentrations of **30** (0, 1.25, 2.5, 5, 10, 20, 40 and 80 µM). ACB downregulated the nuclear pool of phospho-STAT3 pointing out that **30** suppresses the phosphorylation of STAT3, thereby translocating into the nucleus and reducing its DNA binding ability. LSD1 was exploited as nuclear marker and loading control for nuclear protein. The levels of LSD1, cytoplasmic STAT3 and $\beta$-actin did not vary.

These data were also confirmed by in silico studies that determined the acetyl group on the core carbazole as the most auspicious substituent. In particular, in different matched molecular series, compounds bearing the acetyl group showed the best cellular activities. An ethyl substituent is related to a modest solubility of the corresponding compounds, particularly in relation to chlorine substitutions.

On the contrary, they explored *N*-substituents and found the 3-(2-cyano-phenyl)-benzyl substitution as decoration with the highest effects in two out of three matched molecular series. Moreover, they extracted the SH2 domain from the crystal structure of the STAT3 homo-dimer and chose the pTyr as the center for in silico docking experiments. The 3-(2-cyano-phenyl)-benzyl substituted compounds such as **30**, **31**, and **32** demonstrated distinct features in molecular docking apart from conserved shape fit and hydrophobic interactions. The substituents on the carbazole moiety were buried in the pTyr binding site and thereby form hydrogen bonds via the acetyl and hydroxyethyl group. This provides a clarification of the better activity of the acetyl and ethyl decoration over the ethyl one, which misses the acceptor functionality resembling the pTyr in binding to the side chain of Arg609.

Diaz et al. in 2011 [118] explored the possible activity of Lestaurtinib against HL based on the knowledge that JAK/STAT pathway constitutive activation is involved in Hodgkin's lymphoma (HL) pathogenesis and the ability of Lestaurtinib (formerly known as CEP-701, 33 Figure 15) to inhibit JAK2 and FLT3 [119] in myeloproliferative disorders. Genomic gains of JAK2 [120], due to 9p24 gains [121], and SOCS1, a negative regulator of JAK/STAT signaling, are often somatically mutated and inactivated in HL [122]. Besides, constitutive activation of STAT3 has been documented in HL cell lines [123].

To verify, they treated five HL cell lines from refractory patients, L-428, L-1236, L-540, HDML-2, and HD-MY-Z with Lestaurtinib. They noticed a dose-dependent cell growth suppression (23%–66% at 300 nM) and an apoptotic increase (10%–64% at 300 nM) after treatment for 48 h. Furthermore, Lestaurtinib prevented JAK2, STAT5, and STAT3 phosphorylation and decreased the mRNA expression of its downstream antiapoptotic target Bcl-xL. Furthermore, they determined after 1 h, phospho-JAK2 levels were reduced in all the HL cell lines by 46–94% at 300 nM, whereas no substantial changes were detected in JAK2 total protein expression.

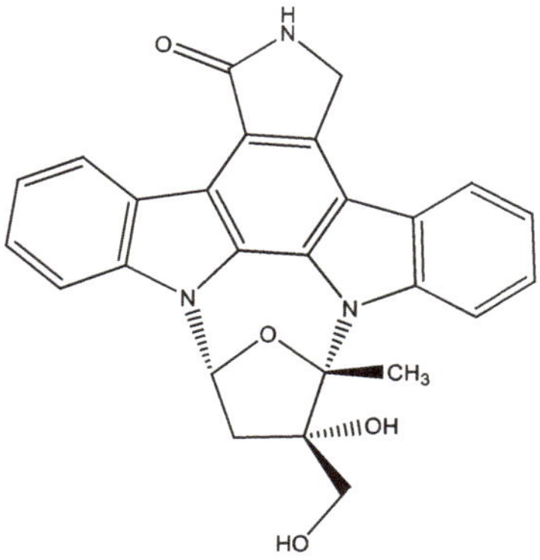

**Figure 15.** Structure of lestaurtinib (CEP-70, **33**).

Lestaurtinib markedly hampered the phosphorylation of STAT5 and STAT3. However, no substantial changes in STAT5 and STAT3 total protein were assessed. Phospho-STAT5 and phosphoSTAT3 levels declined by 88–100% and by 97–100%, respectively, after 1 h of 300 nM of Lestaurtinib treatment. Bcl-xL, a pro-survival protein triggered by phosphorylated STAT5 DNA binding, is upregulated in HL samples, and is implicated in apoptotic resistance in HRS cells. Reduced phosphorylation of STAT5 resulted in lowered mRNA expression of its downstream antiapoptotic effector Bcl-xL. Cells were treated for 1 h with Lestaurtinib at 300 nM discovering that Bcl-xL mRNA expression levels diminished by 52% in L-428, 28% in L-1236, 37% in L540, 55% in HDLM-2, and 71% in HD-MY-Z [124].

Subsequently, Santos et al. [125] carried out a phase 2 study of CEP-701 in which 22 JAK$^{2V617F}$-positive patients with myelofibrosis, for which few therapeutic options are available today, were treated with 80 mg, orally twice daily, of this drug. The total response rate for the patients was of 27%, with a median time to response of three months. Phosphorylated STAT3 levels decreased from baseline in responders while on therapy. A decline in spleen size occurred in three patients (50%). Two patients reached transfusion independency (9%). A dose decrease was necessary for 27% of patients. to reduce toxicity. Hematologic side effects such as anemia and thrombocytopenia were experienced by 14% and 23% of patients, respectively. Nonhematologic toxicity, including diarrhea (72%), nausea (50%), and vomiting (27%), were noted [125].

## 4. Conclusions

The academic community has extensively explored, over the years, heterocyclic compounds that represent more than half of the organic molecules known to exist. Among them a main role is played by carbazoles. This structural motif is found in different naturally occurring alkaloids. An ascending interest in these versatile compounds led both to the development of many synthetic strategies and their extraction from several plants. Indeed, they found an application both in science materials and the pharmaceutical field. In particular, some prior studies suggest that they could be promising anticancer agents. [126] According to some studies their effects could be due to the involvement of the JAK/STAT pathway. Signal transducers and activators of transcription (STATs) are a family of cytoplasmic proteins with roles like signal messengers and transcription factors engaged in normal cellular responses to cytokines and growth factors. Associated with STATs is the Janus kinases (JAK), belonging to the tyrosine kinases family, responsible for activating the

STAT cascade that initiates, ultimately, gene transcription. [127] Literature data report that carbazoles (Table 1) could act by downregulating STAT proteins (particularly STAT-3), and also affecting interleukins and i-NOS production.

**Table 1.** Carbazole derivatives as stat inhibitors.

| Compound | Name Compound | Structure Compound | References |
|---|---|---|---|
| **1** | Carbazole | 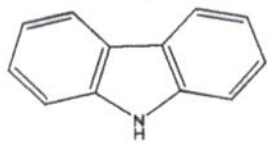 | Arbiser et al. [31] |
| **11** | 2-Hydroxycarbazole | 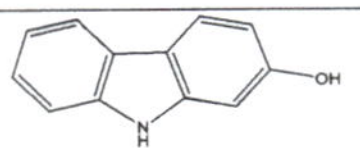 | Arbiser et al. [31] |
| **12** | Mahanine 3,11-dihydro-3,5-dimethyl-3-(4-methyl-3-pentenyl)-pyrano[3,2-*a*]carbazol-9-ol | 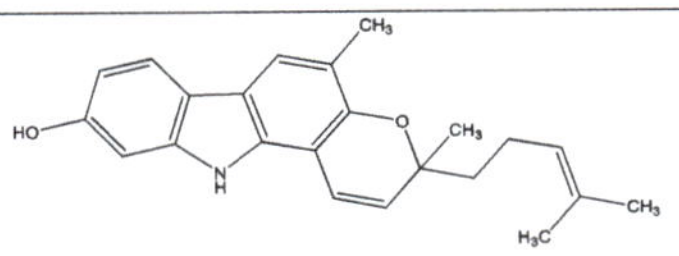 | Sarkar et al. [100] |
| **13a** | Dimethyl-5-(5-(6-methoxy-1,4-dimethyl-9*H*-carbazol-9-yl) pentyloxy)isophthalate | 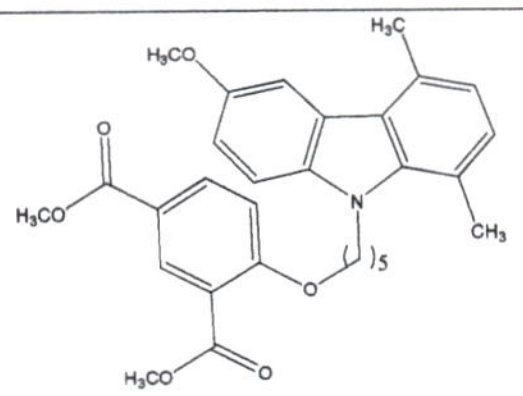 | Saturnino et al. [32] |
| **13b** | Dimethyl-5-(6-(6-methoxy-1,4-dimethyl-9*H*-carbazol-9-yl) hexyloxy)isophthalate | 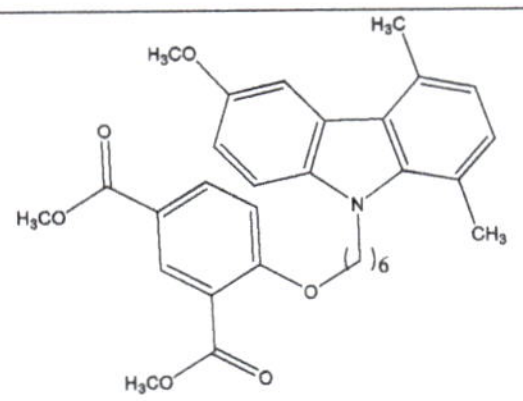 | Saturnino et al. [32] |
| **13c** | Dimethyl-5-(7-(6-methoxy-1,4-dimethyl-9*H*-carbazol-9-yl) heptyloxy)isophthalate | 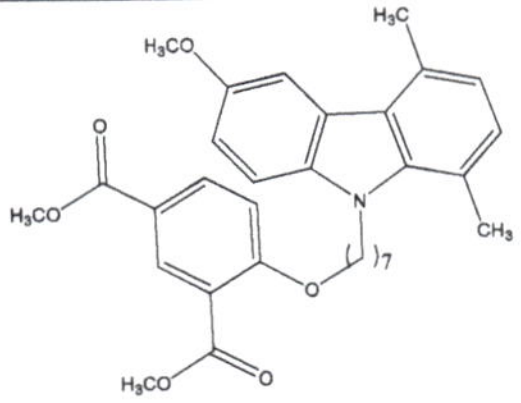 | Saturnino et al. [32] |
| **14a** | Methyl-2-(5-(6-methoxy-1,4-dimethyl-9*H*-carbazol-9-yl) pentyloxy) benzoate | 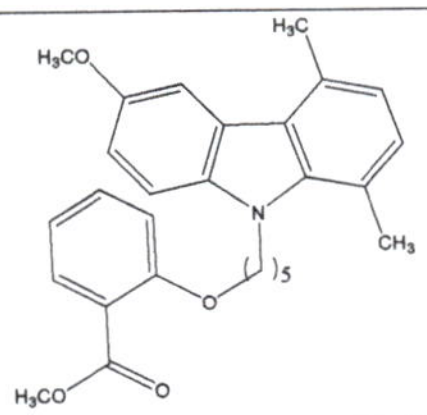 | Saturnino et al. [32] |

**Table 1.** *Cont.*

| Compound | Name Compound | Structure Compound | References |
|---|---|---|---|
| **14b** | Methyl-2-(6-(6-methoxy-1,4-dimethyl-9*H*-carbazol-9-yl)hexyloxy)benzoate | | Saturnino et al. [32] |
| **14c** | Methyl-2-(7-(6-methoxy-1,4-dimethyl-9*H*-carbazol-9-yl)heptyloxy)benzoate | | Saturnino et al. [32] |
| **15** | 7-Hydroxy-1-methyl-9*H*-carbazol-2-yl-5-(dimethylamino)-naphthalene-1-sulfonate | | Hou et al. [107]; Kang et al. [111] |
| **16** | EC-70124 (5*R*,7*S*,9*S*)-7,8-Dihydroxy-9-methyl-6,7,8,9-tetrahydro-5*H*,14*H*-17-oxa-4b,9a,15-triaza-5,9-methanodibenzo[b,h]cyclonona[jkl]cyclopenta[e]-as-indacene-14,16(15*H*)-dione | | Cuenca-López et al. [113] Estupiñan et al. [114] |
| **17** | 1,4-Dimethyl-9*H*-carbazole | | Botta et al. [115] |
| **18** | 6-Hydroxy-1,4-dimethyl-9*H*-carbazole | | Botta et al. [115] |
| **19** | 6-Methoxy-1,4-dimethyl-9*H*-carbazole | | Botta et al. [115] |
| **20** | Ethyl-5,8-dimethyl-9*H*-carbazole-3-carboxylate | | Botta et al. [115] |

**Table 1.** *Cont.*

| Compound | Name Compound | Structure Compound | References |
|---|---|---|---|
| 21 | 6-Chloro-1,4-dimethyl-9*H*-carbazole | 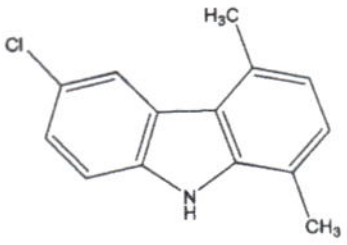 | Botta et al. [115] |
| 22 | 5,8-Dimethyl-9*H*-carbazole-3-sulfonamide | 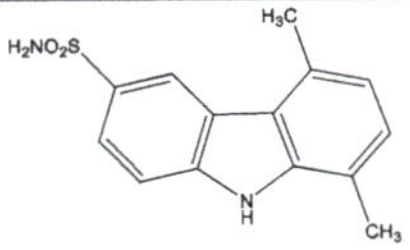 | Botta et al. [115] |
| 23 | 6-Methoxy-1,4-dimethyl-3-nitro-9*H*-carbazole | 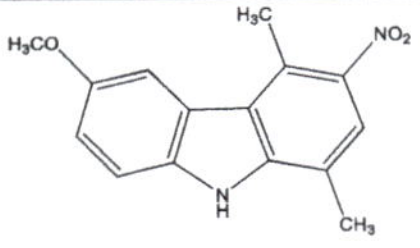 | Botta et al. [115] |
| 24 | Ethyl-5,8-dimethyl-9*H*-carbazole-3-nitro-carboxylate | 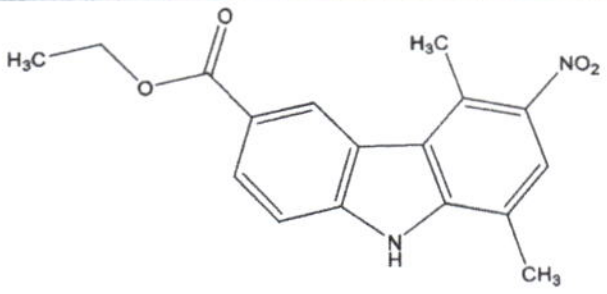 | Botta et al. [115] |
| 25 | 6-Methoxy-1,4,9-trimethyl-carbazole | 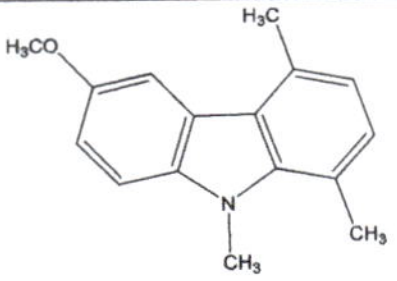 | Botta et al. [115] |
| 26 | 6-Methoxy-1,4-dimethyl-9-ethyl-carbazole | 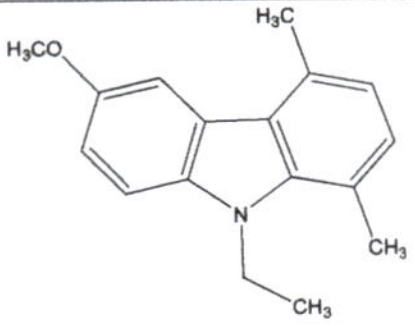 | Botta et al. [115] |
| 27 | 3-(3,4-Dicholophenyl)-6-(morpholine-4-carbonyl)-9*H*-carbazole-1-carboxamide | 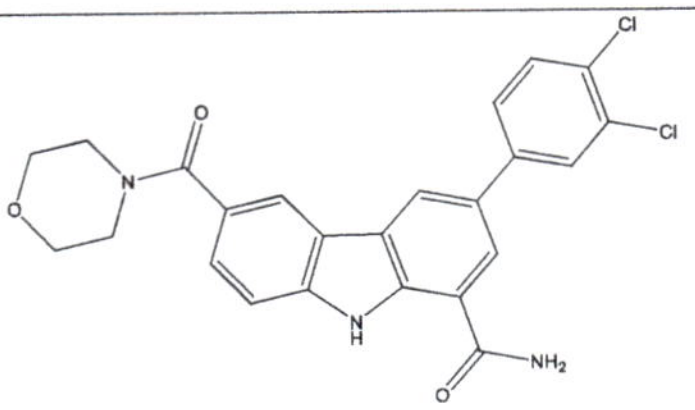 | Zimmermann et al. [116] |
| 28 | (*S*)-7-(2-(dimethylamino)-pyrrolidine-1-carbonyl)-3-(3-methoxyphenyl)-9*H*-carbazole-1-carboxamide | 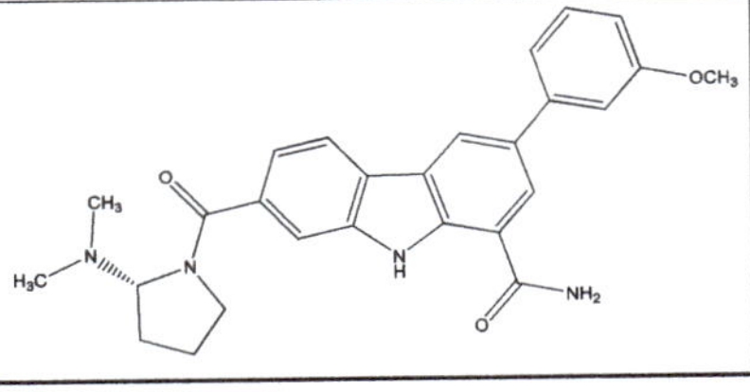 | Zimmermann et al. [116] |

**Table 1.** *Cont.*

| Compound | Name Compound | Structure Compound | References |
|---|---|---|---|
| **29** | 3-(1-Methyl-1*H*-indazol-5-yl)-7-(4-methylpiperazin-1-yl)-9*H*-carbazole-1-carboxamide | 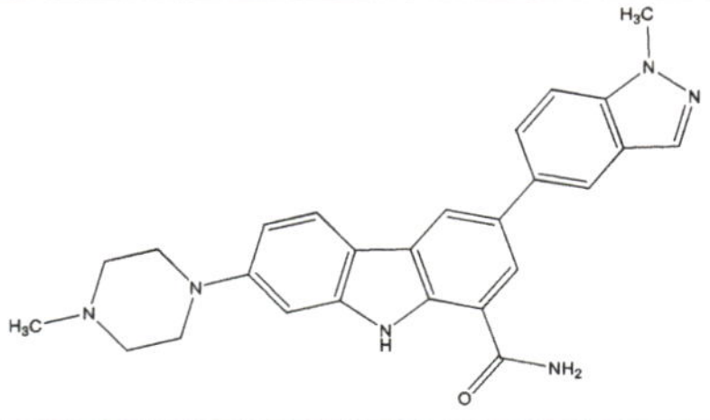 | Zimmermann et al. [116] |
| **30** | 3′-((3-Acetyl-6-chloro-9*H*-carbazol-9-yl)methyl)-[1,1′-biphenyl]-2-carbonitrile | 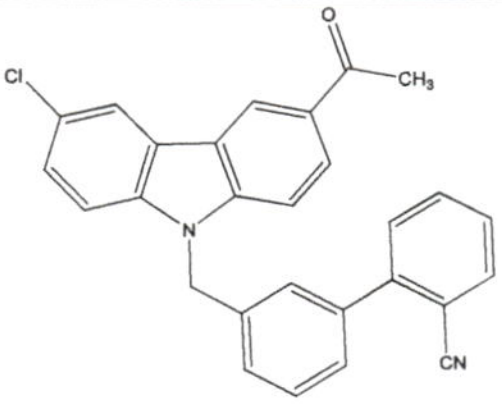 | Baburajeev et al. [117] |
| **31** | 3′-((3-Chloro-6-ethyl-9*H*-carbazol-9-yl)methyl)-[1,1′-biphenyl]-2-carbonitrile | 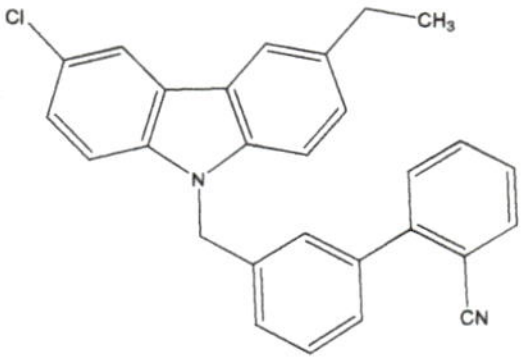 | Baburajeev et al. [117] |
| **32** | 3′-((3-Chloro-6-(1-hydroxyethyl)-9*H*-carbazol-9-yl)methyl)-[1,1′-biphenyl]-2-carbonitrile | 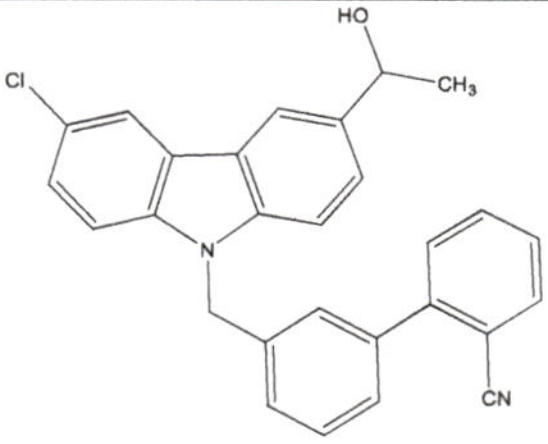 | Baburajeev et al. [117] |
| **33** | CEP-701 Lestaurtinib (5*S*,6*S*,8*R*)-6-Hydroxy-6-(hydroxymethyl)-5-methyl-7,8,14,15-tetrahydro-5*H*-16-oxa-4b,8a,14-triaza-5,8-methanodibenzo[b,h]cycloocta[jkl]cyclopenta[e]-as-indacen-13(6*H*)-one | 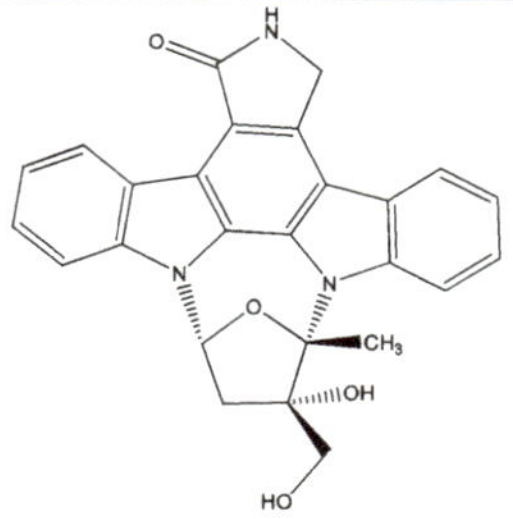 | Diaz et al. [118] Shabbir et al. [119] Geyer et al. [124] Santos et al. [125] |

Figure 16 represents the mechanisms of action of the compounds cited in this review.

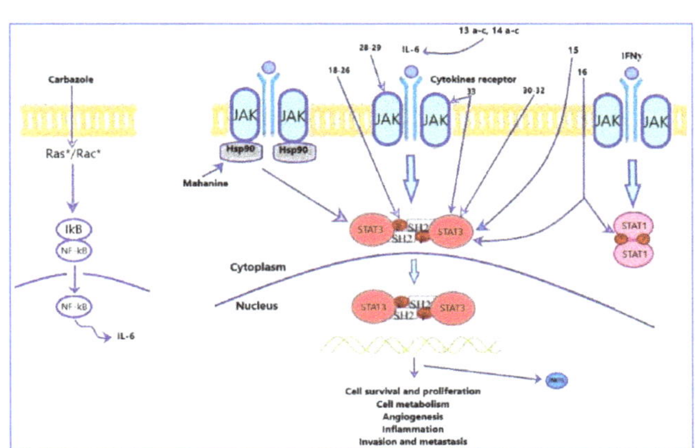

**Figure 16.** Mechanism of action of compounds carbazole mahanine and **13 a–c**, **14 a–c**, **15**, **16**, **18–26**, **28**, **29**, **30–33**.

Some of these studies have also been supported by in silico evaluations. It may be interesting to compare the activity between some successful carbazole and non-carbazole molecules targeting STATs. Indeed, HJC0123, cpd30, cpd188, and carbazolic compound 15 demonstrated their ability to target STAT3 and to exert antitumoral effects against breast cancer. Among them, HJC0123 exhibited the lower $IC_{50}$ values, amounting to 0.1–1.25 µM. Furthermore, it demonstrated to be orally bioavailable and suppressed tumoral growth in vivo [90]. Concerning compound **15**, its mechanism of action on STAT3 has been deepened, demonstrating its ability to inhibit phospho-STAT3 in the micromolar range. Moreover, other findings also reported interesting activity in an in vivo model of breast cancer in which the tumor suppression occurred after a few days of treatment. All this data, taken together, make compound **15** a promising carbazolic lead compound and a potential tool in the fight against cancer. However, further studies able to validate the importance of compound **15** and of other promising carbazoles for clinical use must be carried out. If they succeed, carbazoles could represent a valid alternative to conventional treatments, allowing us to overcome the phenomena of drug resistance, entering, in this way, into the plethora of target therapies.

**Author Contributions:** Conceptualization, A.C. (Anna Caruso) and C.S.; methodology, A.B.; software, G.S.; resources, A.C. (Anna Caruso) and A.B.; data curation, A.C. (Alessia Carocci); writing—original draft preparation, A.C. (Anna Caruso) and A.B.; writing—review and editing, A.C. (Alessia Carocci); supervision, M.S.S. and C.S. All authors have read and agreed to the published version of the manuscript.

**Funding:** This research received no external funding.

**Institutional Review Board Statement:** Not applicable.

**Informed Consent Statement:** Not applicable.

**Data Availability Statement:** Not applicable.

**Conflicts of Interest:** The authors declare no conflict of interest.

## References

1. Chin, Y.W.; Balunas, M.J.; Chai, H.B.; Kinghorn, A.D. Drug discovery from natural sources. *AAPS J.* **2006**, *8*, E239–E253. [CrossRef]
2. Cordell, G.A.; Quinn-Beattie, M.L.; Farnsworth, N.R. The potential of alkaloids in drug discovery. *Phytother Res.* **2001**, *15*, 183–205. [CrossRef]

3. Bhambhani, S.; Kondhare, K.R.; Giri, A.P. Diversity in chemical structures and biological properties of plant alkaloids. *Molecules* **2021**, *26*, 3374. [CrossRef]
4. Koehn, F.E.; Carter, G.T. The evolving role of natural products in drug discovery. *Nat. Rev. Drug Discov.* **2005**, *4*, 206–220. [CrossRef]
5. Mital, A. Synthetic nitroimidazoles: Biological activities and mutagenicity relationships. *Sci. Pharm.* **2009**, *77*, 497–520. [CrossRef]
6. Nekrasov, D. Biological activity of 5-and 6-membered azaheterocycles and their synthesis from 5-aryl-2, 3-dihydrofuran-2, 3-diones. *Chem. Heterocycl. Compd.* **2001**, *37*, 263–275. [CrossRef]
7. Laurence, C.; Brameld, K.A.; Graton, J.; Le Questel, J.-Y.; Renault, E. The p K BHX database: Toward a better understanding of hydrogen-bond basicity for medicinal chemists. *J. Med. Chem.* **2009**, *52*, 4073–4086. [CrossRef] [PubMed]
8. Graebe, C.; Glaser, C. Ueber carbazol. *Justus Liebigs Ann. Chem.* **1872**, *163*, 343–360. [CrossRef]
9. Chakraborty, D.P.; Barman, B.K.; Bose, P.K. On the constitution of murrayanine, a carbazole derivative isolated from Murraya koenigii Spreng. *Tetrahedron* **1965**, *21*, 681–685. [CrossRef]
10. Saturnino, C.; Caruso, A.; Longo, P.; Capasso, A.; Pingitore, A.; Cristina Caroleo, M.; Cione, E.; Perri, M.; Nicolò, F.; Mollica Nardo, V. Crystallographic study and biological evaluation of 1, 4-dimethyl-N-alkylcarbazoles. *Curr. Top. Med. Chem.* **2015**, *15*, 973–979. [CrossRef]
11. Panno, A.; Sinicropi, M.S.; Caruso, A.; El-Kashef, H.; Lancelot, J.C.; Aubert, G.; Lesnard, A.; Cresteil, T.; Rault, S. New trimethoxybenzamides and trimethoxyphenylureas derived from dimethylcarbazole as cytotoxic agents. Part, I. *J. Heterocycl. Chem.* **2014**, *51*, E294–E302. [CrossRef]
12. Caruso, A.; Sinicropi, M.S.; Lancelot, J.-C.; El-Kashef, H.; Saturnino, C.; Aubert, G.; Ballandonne, C.; Lesnard, A.; Cresteil, T.; Dallemagne, P. Synthesis and evaluation of cytotoxic activities of new guanidines derived from carbazoles. *Bioorg. Med. Chem. Lett.* **2014**, *24*, 467–472. [CrossRef] [PubMed]
13. Sinicropi, M.S.; Iacopetta, D.; Rosano, C.; Randino, R.; Caruso, A.; Saturnino, C.; Muià, N.; Ceramella, J.; Puoci, F.; Rodriquez, M. N-thioalkylcarbazoles derivatives as new anti-proliferative agents: Synthesis, characterisation and molecular mechanism evaluation. *J. Enzyme Inhib. Med. Chem.* **2018**, *33*, 434–444. [CrossRef] [PubMed]
14. Knölker, H.-J.; Reddy, K.R. Isolation and synthesis of biologically active carbazole alkaloids. *Chem. Rev.* **2002**, *102*, 4303–4428. [CrossRef]
15. Bashir, M.; Bano, A.; Ijaz, A.S.; Chaudhary, B.A. Recent developments and biological activities of N-substituted carbazole derivatives: A review. *Molecules* **2015**, *20*, 13496–13517. [CrossRef] [PubMed]
16. Is, O.D.; Koyuncu, F.B.; Koyuncu, S.; Ozdemir, E. A new imine coupled pyrrole–carbazole–pyrrole polymer: Electro-optical properties and electrochromism. *Polymer* **2010**, *51*, 1663–1669. [CrossRef]
17. Srinivas, K.; Kumar, C.R.; Reddy, M.A.; Bhanuprakash, K.; Rao, V.J.; Giribabu, L. D-π-A organic dyes with carbazole as donor for dye-sensitized solar cells. *Synth. Met.* **2011**, *161*, 96–105. [CrossRef]
18. Lee, K.S.; Lee, D.U.; Choo, D.C.; Kim, T.W.; Ryu, E.D.; Kim, S.W.; Lim, J.S. Organic light-emitting devices fabricated utilizing core/shell CdSe/ZnS quantum dots embedded in a polyvinylcarbazole. *J. Mater. Sci.* **2011**, *46*, 1239–1243. [CrossRef]
19. Wang, L.; Fu, Y.; Zhu, L.; Cui, G.; Liang, F.; Guo, L.; Zhang, X.; Xie, Z.; Su, Z. Synthesis and photovoltaic properties of low-bandgap polymers based on N-arylcarbazole. *Polymer* **2011**, *52*, 1748–1754. [CrossRef]
20. Adhikari, R.M.; Neckers, D.C.; Shah, B.K. Photophysical study of blue, green, and orange-red light-emitting carbazoles. *J. Org. Chem.* **2009**, *74*, 3341–3349. [CrossRef]
21. Caruso, A.; Ceramella, J.; Iacopetta, D.; Saturnino, C.; Mauro, M.V.; Bruno, R.; Aquaro, S.; Sinicropi, M.S. Carbazole derivatives as antiviral agents: An overview. *Molecules* **2019**, *24*, 1912. [CrossRef]
22. Caruso, A.; Iacopetta, D.; Puoci, F.; Rita Cappello, A.; Saturnino, C.; Stefania Sinicropi, M. Carbazole derivatives: A promising scenario for breast cancer treatment. *Mini Rev. Med. Chem* **2016**, *16*, 630–643. [CrossRef]
23. Roy, J.; Jana, A.K.; Mal, D. Recent trends in the synthesis of carbazoles: An update. *Tetrahedron* **2012**, *68*, 6099–6121. [CrossRef]
24. Caruso, A.; Voisin-Chiret, A.S.; Lancelot, J.-C.; Sinicropi, M.S.; Garofalo, A.; Rault, S. Novel and efficient synthesis of 5, 8-dimethyl-9H-carbazol-3-ol via a hydroxydeboronation reaction. *Heterocycles* **2007**, *71*, 2203–2210. [CrossRef]
25. Caruso, A.; Voisin-Chiret, A.S.; Lancelot, J.-C.; Sinicropi, M.S.; Garofalo, A.; Rault, S. Efficient and simple synthesis of 6-Aryl-1, 4-dimethyl-9H-carbazoles. *Molecules* **2008**, *13*, 1312–1320. [CrossRef]
26. Saturnino, C.; Grande, F.; Aquaro, S.; Caruso, A.; Iacopetta, D.; Bonomo, M.G.; Longo, P.; Schols, D.; Sinicropi, M.S. Chloro-1, 4-dimethyl-9H-carbazole derivatives displaying anti-HIV activity. *Molecules* **2018**, *23*, 286. [CrossRef]
27. You, X.; Zhu, D.; Lu, W.; Sun, Y.; Qiao, S.; Luo, B.; Du, Y.; Pi, R.; Hu, Y.; Huang, P. Design, synthesis and biological evaluation of N-arylsulfonyl carbazoles as novel anticancer agents. *RSC Adv.* **2018**, *8*, 17183–17190. [CrossRef]
28. Ceramella, J.; Iacopetta, D.; Barbarossa, A.; Caruso, A.; Grande, F.; Bonomo, M.G.; Mariconda, A.; Longo, P.; Carmela, S.; Sinicropi, M.S. Carbazole derivatives as kinase-targeting inhibitors for cancer treatment. *Mini Rev. Med. Chem* **2020**, *20*, 444–465. [CrossRef] [PubMed]
29. Ceramella, J.; Caruso, A.; Occhiuzzi, M.A.; Iacopetta, D.; Barbarossa, A.; Rizzuti, B.; Dallemagne, P.; Rault, S.; El-Kashef, H.; Saturnino, C. Benzothienoquinazolinones as new multi-target scaffolds: Dual inhibition of human Topoisomerase I and tubulin polymerization. *Eur. J. Med. Chem.* **2019**, *181*, 111583. [CrossRef] [PubMed]

30. Saturnino, C.; Caruso, A.; Iacopetta, D.; Rosano, C.; Ceramella, J.; Muià, N.; Mariconda, A.; Bonomo, M.G.; Ponassi, M.; Rosace, G. Inhibition of human topoisomerase II by N, N, N-trimethylethanammonium iodide alkylcarbazole derivatives. *Chem. Med. Chem.* **2018**, *13*, 2635–2643. [CrossRef]

31. Arbiser, J.L.; Govindarajan, B.; Battle, T.E.; Lynch, R.; Frank, D.A.; Ushio-Fukai, M.; Perry, B.N.; Stern, D.F.; Bowden, G.T.; Liu, A. Carbazole is a naturally occurring inhibitor of angiogenesis and inflammation isolated from antipsoriatic coal tar. *J. Invest. Dermatol.* **2006**, *126*, 1396–1402. [CrossRef]

32. Saturnino, C.; Palladino, C.; Napoli, M.; Sinicropi, M.S.; Botta, A.; Sala, M.; de Prati, A.C.; Novellino, E.; Suzuki, H. Synthesis and biological evaluation of new N-alkylcarbazole derivatives as STAT3 inhibitors: Preliminary study. *Eur. J. Med. Chem.* **2013**, *60*, 112–119. [CrossRef]

33. Akira, S. Roles of STAT3 defined by tissue-specific gene targeting. *Oncogene* **2000**, *19*, 2607–2611. [CrossRef]

34. Smithgall, T.E.; Briggs, S.D.; Schreiner, S.; Lerner, E.C.; Cheng, H.; Wilson, M.B. Control of myeloid differentiation and survival by Stats. *Oncogene* **2000**, *19*, 2612–2618. [CrossRef]

35. Villarino, A.V.; Kanno, Y.; O'Shea, J.J. Mechanisms and consequences of Jak–STAT signaling in the immune system. *Nat. Immunol.* **2017**, *18*, 374–384. [CrossRef] [PubMed]

36. Lee, B.; Gopal, R.; Manni, M.L.; McHugh, K.J.; Mandalapu, S.; Robinson, K.M.; Alcorn, J.F. STAT1 is required for suppression of type 17 immunity during influenza and bacterial superinfection. *Immuno. Horizons* **2017**, *1*, 81–91. [CrossRef]

37. Machiraju, D.; Moll, I.; Gebhardt, C.; Sucker, A.; Buder-Bakhaya, K.; Schadendorf, D.; Hassel, J.C. STAT5 expression correlates with recurrence and survival in melanoma patients treated with interferon-$\alpha$. *Melanoma Res.* **2018**, *28*, 204–210. [CrossRef] [PubMed]

38. Gamero, A.M.; Young, M.R.; Mentor-Marcel, R.; Bobe, G.; Scarzello, A.J.; Wise, J.; Colburn, N.H. STAT2 contributes to promotion of colorectal and skin carcinogenesis. *Cancer Prev. Res.* **2010**, *3*, 495–504. [CrossRef] [PubMed]

39. Akira, S. Functional roles of STAT family proteins: Lessons from knockout mice. *Stem Cells.* **1999**, *17*, 138–146. [CrossRef] [PubMed]

40. Macias, E.; Rao, D.; DiGiovanni, J. Role of stat3 in skin carcinogenesis: Insights gained from relevant mouse models. *J. Skin Cancer* **2013**, *2013*, 684050. [CrossRef]

41. Darnell, J.E. Validating stat3 in cancer therapy. *Nat. Med.* **2005**, *11*, 595–596. [CrossRef]

42. Tweardy, D.J.; Jing, N. Enhancing or eliminating signals for cell survival to treat disease. *Trans. Am. Clin. Climatol. Assoc.* **2006**, *117*, 33.

43. Buettner, R.; Mora, L.B.; Jove, R. Activated STAT signaling in human tumors provides novel molecular targets for therapeutic intervention. *Clin. Cancer Res.* **2002**, *8*, 945–954.

44. Gao, P.; Niu, N.; Wei, T.; Tozawa, H.; Chen, X.; Zhang, C.; Zhang, J.; Wada, Y.; Kapron, C.M.; Liu, J. The roles of signal transducer and activator of transcription factor 3 in tumor angiogenesis. *Oncotarget* **2017**, *8*, 69139–69161. [CrossRef]

45. Dutta, P.; Zhang, L.; Zhang, H.; Peng, Q.; Montgrain, P.R.; Wang, Y.; Song, Y.; Li, J.; Li, W.X. Unphosphorylated STAT3 in heterochromatin formation and tumor suppression in lung cancer. *BMC Cancer* **2020**, *20*, 145. [CrossRef]

46. Loh, C.Y.; Arya, A.; Naema, A.F.; Wong, W.F.; Sethi, G.; Looi, C.Y. Signal transducer and activator of transcription (STATs) proteins in cancer and inflammation: Functions and therapeutic implication. *Front. Oncol.* **2019**, *9*, 48. [CrossRef]

47. Andersson, E.I.; Brück, O.; Braun, T.; Mannisto, S.; Saikko, L.; Lagström, S.; Ellonen, P.; Leppä, S.; Herling, M.; Kovanen, P.E.; et al. STAT3 mutation is associated with STAT3 activation in CD30$^+$ ALK$^-$ ALCL. *Cancers* **2020**, *12*, 702. [CrossRef]

48. Szabo, S.J.; Sullivan, B.M.; Peng, S.L.; Glimcher, L.H. Molecular mechanisms regulating Th1 immune responses. *Annu. Rev. Immunol.* **2003**, *21*, 713–758. [CrossRef]

49. Lovett-Racke, A.E.; Yang, Y.; Racke, M.K. Th1 versus Th17: Are T cell cytokines relevant in multiple sclerosis? *Biochim. Biophys. Acta Mol. Basis Dis.* **2011**, *1812*, 246–251. [CrossRef]

50. Lin, J.-X.; Leonard, W.J. The role of stat5a and stat5b in signaling by IL-2 family cytokines. *Oncogene* **2000**, *19*, 2566–2576. [CrossRef]

51. Buitenhuis, M.; Baltus, B.; Lammers, J.-W.J.; Coffer, P.J.; Koenderman, L. Signal transducer and activator of transcription 5a (STAT5a) is required for eosinophil differentiation of human cord blood–derived CD34$^+$ cells. *Blood* **2003**, *101*, 134–142. [CrossRef]

52. Nelson, E.A.; Walker, S.R.; Weisberg, E.; Bar-Natan, M.; Barrett, R.; Gashin, L.B.; Terrell, S.; Klitgaard, J.L.; Santo, L.; Addorio, M.R. The STAT5 inhibitor pimozide decreases survival of chronic myelogenous leukemia cells resistant to kinase inhibitors. *Blood* **2011**, *117*, 3421–3429. [CrossRef]

53. Cotarla, I.; Ren, S.; Zhang, Y.; Gehan, E.; Singh, B.; Furth, P.A. Stat5a is tyrosine phosphorylated and nuclear localized in a high proportion of human breast cancers. *Int. J. Cancer* **2004**, *108*, 665–671. [CrossRef] [PubMed]

54. Benekli, M.; Baumann, H.; Wetzler, M. Targeting signal transducer and activator of transcription signaling pathway in leukemias. *J. Clin. Oncol.* **2009**, *27*, 4422. [CrossRef]

55. Nelson, E.A.; Sharma, S.V.; Settleman, J.; Frank, D.A. A chemical biology approach to developing STAT inhibitors: Molecular strategies for accelerating clinical translation. *Oncotarget* **2011**, *2*, 518. [CrossRef] [PubMed]

56. Kaymaz, B.T.; Selvi, N.; Gokbulut, A.A.; Aktan, C.; Gündüz, C.; Saydam, G.; Sahin, F.; Cetintaş, V.B.; Baran, Y.; Kosova, B. Suppression of STAT5A and STAT5B chronic myeloid leukemia cells via siRNA and antisense-oligonucleotide applications with the induction of apoptosis. *Am. J. Blood Res.* **2013**, *3*, 58–70.

57. Sehra, S.; Yao, Y.; Howell, M.D.; Nguyen, E.T.; Kansas, G.S.; Leung, D.Y.; Travers, J.B.; Kaplan, M.H. IL-4 regulates skin homeostasis and the predisposition toward allergic skin inflammation. *J. Immunol.* **2010**, *184*, 3186–3190. [CrossRef]

58. Chapoval, S.P.; Dasgupta, P.; Smith, E.P.; DeTolla, L.J.; Lipsky, M.M.; Kelly-Welch, A.E.; Keegan, A.D. STAT6 expression in multiple cell types mediates the cooperative development of allergic airway disease. *J. Immunol.* **2011**, *186*, 2571–2583. [CrossRef] [PubMed]

59. Chiba, Y.; Nakazawa, S.; Todoroki, M.; Shinozaki, K.; Sakai, H.; Misawa, M. Interleukin-13 augments bronchial smooth muscle contractility with an up-regulation of RhoA protein. *Am. J. Respir. Cell Mol. Biol.* **2009**, *40*, 159–167. [CrossRef]

60. Chiba, Y.; Todoroki, M.; Nishida, Y.; Tanabe, M.; Misawa, M. A novel STAT6 inhibitor AS1517499 ameliorates antigen-induced bronchial hypercontractility in mice. *Am. J. Respir. Cell Mol. Biol.* **2009**, *41*, 516–524. [CrossRef]

61. Khaled, W.T.; Read, E.K.; Nicholson, S.E.; Baxter, F.O.; Brennan, A.J.; Came, P.J.; Sprigg, N.; McKenzie, A.N.; Watson, C.J. The IL-4/IL-13/Stat6 signalling pathway promotes luminal mammary epithelial cell development. *Development* **2007**, *134*, 2739–2750. [CrossRef] [PubMed]

62. Chen, J.; Gong, C.; Mao, H.; Li, Z.; Fang, Z.; Chen, Q.; Lin, M.; Jiang, X.; Hu, Y.; Wang, W.; et al. E2F1/SP3/STAT6 axis is required for IL-4-induced epithelial-mesenchymal transition of colorectal cancer cells. *Int. J. Oncol.* **2018**, *53*, 567–578. [CrossRef]

63. Rawlings, J.S.; Rosler, K.M.; Harrison, D.A. The JAK/STAT signaling pathway. *J. Cell Sci.* **2004**, *117*, 1281–1283. [CrossRef] [PubMed]

64. Sarid, N.; Eshel, R.; Rahamim, E.; Carmiel, M.; Kirgner, I.; Shpringer, M.; Trestman, S.; Marilus, R.; Perry, C.; Polliack, A.; et al. JAK2 mutation: An aid in the diagnosis of occult myeloproliferative neoplasms in patients with major intraabdominal vein thrombosis and normal blood counts. *Isr. Med. Assoc. J.* **2013**, *15*, 698–700. [PubMed]

65. López-Matencio, J.M.S.; Baladrón, A.M.; Castañeda, S. JAK-STAT inhibitors for the treatment of immunomediated diseases. *Med. Clin.* **2019**, *152*, 353–360. [CrossRef]

66. Miklossy, G.; Hilliard, T.S.; Turkson, J. Therapeutic modulators of STAT signalling for human diseases. *Nat. Rev. Drug Discov.* **2013**, *12*, 611–629. [CrossRef] [PubMed]

67. Bromberg, J.F.; Horvath, C.M.; Besser, D.; Lathem, W.W.; Darnell, J.E. Stat3 activation is required for cellular transformation by v-src. *Mol. Cell. Biol.* **1998**, *18*, 2553–2558. [CrossRef]

68. Turkson, J.; Bowman, T.; Garcia, R.; Caldenhoven, E.; De Groot, R.P.; Jove, R. Stat3 activation by Src induces specific gene regulation and is required for cell transformation. *Mol. Cell. Biol.* **1998**, *18*, 2545–2552. [CrossRef]

69. Bromberg, J.F.; Wrzeszczynska, M.H.; Devgan, G.; Zhao, Y.; Pestell, R.G.; Albanese, C.; Darnell, J.E., Jr. Stat3 as an oncogene. *Cell* **1999**, *98*, 295–303. [CrossRef]

70. Zong, C.S.; Zeng, L.; Jiang, Y.; Sadowski, H.B.; Wang, L.-H. Stat3 plays an important role in oncogenic Ros-and insulin-like growth factor I receptor-induced anchorage-independent growth. *J. Biol. Chem.* **1998**, *273*, 28065–28072. [CrossRef]

71. Zhang, N.; Zheng, Q.; Wang, Y.; Lin, J.; Wang, H.; Liu, R.; Yan, M.; Chen, X.; Yang, J.; Chen, X. Renoprotective effect of the recombinant anti-IL-6R fusion proteins by inhibiting JAK2/STAT3 signaling pathway in diabetic nephropathy. *Front. Pharmacol.* **2021**, *12*, 681424. [CrossRef]

72. Wen, X.; Lin, H.H.; Shih, H.-M.; Kung, H.-J.; Ann, D.K. Kinase activation of the non-receptor tyrosine kinase Etk/BMX alone is sufficient to transactivate STAT-mediated gene expression in salivary and lung epithelial cells. *J. Biol. Chem.* **1999**, *274*, 38204–38210. [CrossRef] [PubMed]

73. Lund, T.; Medveczky, M.M.; Medveczky, P.G. Herpesvirus saimiri Tip-484 membrane protein markedly increases p56lck activity in T cells. *J. Virol.* **1997**, *71*, 378–382. [CrossRef] [PubMed]

74. Migone, T.-S.; Lin, J.X.; Cereseto, A.; Mulloy, J.C.; O'Shea, J.J.; Franchini, G.; Leonard, W.J. Constitutively activated Jak-STAT pathway in T cells transformed with HTLV-I. *Science* **1995**, *269*, 79–81. [CrossRef] [PubMed]

75. Garcia, R.; Yu, C.L.; Hudnall, A.; Catlett, R.; Nelson, K.L.; Smithgall, T.; Fujita, D.J.; Ethier, S.P.; Jove, R. Constitutive activation of Stat3 in fibroblasts transformed by diverse oncoproteins and in breast carcinoma cells. *Cell Growth Differ.* **1997**, *8*, 1267–1276. [PubMed]

76. Weber-Nordt, R.; Egen, C.; Wehinger, J.; Ludwig, W.; Gouilleux-Gruart, V.; Mertelsmann, R.; Finke, J. Constitutive activation of STAT proteins in primary lymphoid and myeloid leukemia cells and in Epstein-Barr virus (EBV)-related lymphoma cell lines. *Blood* **1996**, *88*, 809–816. [CrossRef]

77. Gharibi, T.; Babaloo, Z.; Hosseini, A.; Abdollahpour-alitappeh, M.; Hashemi, V.; Marofi, F.; Nejati, K.; Baradaran, B. Targeting STAT3 in cancer and autoimmune diseases. *Eur. J. Pharmacol.* **2020**, *878*, 173107. [CrossRef]

78. Turkson, J. STAT proteins as novel targets for cancer drug discovery. *Expert Opin. Ther. Targets* **2004**, *8*, 409–422. [CrossRef]

79. Turkson, J.; Ryan, D.; Kim, J.S.; Zhang, Y.; Chen, Z.; Haura, E.; Laudano, A.; Sebti, S.; Hamilton, A.D.; Jove, R. Phosphotyrosyl peptides block Stat3-mediated DNA binding activity, gene regulation, and cell transformation. *J. Biol. Chem.* **2001**, *276*, 45443–45455. [CrossRef] [PubMed]

80. McMurray, J.S.; Ren, Z.; Mandal, P.K.; Chen, X. Model of intermolecular interactions between high affinity phosphopeptides and Stat3. *Adv. Exp. Med. Biol.* **2009**, *611*, 543–544.

81. McMurray, J.S. Structural basis for the binding of high affinity phosphopeptides to Stat3. *Biopolymers* **2008**, *90*, 69–79. [CrossRef] [PubMed]

82. Fletcher, S.; Drewry, J.A.; Shahani, V.M.; Page, B.D.; Gunning, P.T. Molecular disruption of oncogenic signal transducer and activator of transcription 3 (STAT3) protein. *Biochem. Cell Biol.* **2009**, *87*, 825–833. [CrossRef]

83. Zhao, W.; Jaganathan, S.; Turkson, J. A cell-permeable Stat3 SH2 domain mimetic inhibits Stat3 activation and induces antitumor cell effects in vitro. *J. Biol. Chem.* **2010**, *285*, 35855–35865. [CrossRef]
84. McMurray, J.S. A new small-molecule Stat3 inhibitor. *Chem. Biol.* **2006**, *13*, 1123–1124. [CrossRef] [PubMed]
85. Fuh, B.; Sobo, M.; Cen, L.; Josiah, D.; Hutzen, B.; Cisek, K.; Bhasin, D.; Regan, N.; Lin, L.; Chan, C. LLL-3 inhibits STAT3 activity, suppresses glioblastoma cell growth and prolongs survival in a mouse glioblastoma model. *Br. J. Cancer* **2009**, *100*, 106–112. [CrossRef]
86. Hao, W.; Hu, Y.; Niu, C.; Huang, X.; Chang, C.-P.B.; Gibbons, J.; Xu, J. Discovery of the catechol structural moiety as a Stat3 SH2 domain inhibitor by virtual screening. *Bioorg. Med. Chem. Lett.* **2008**, *18*, 4988–4992. [CrossRef]
87. Siddiquee, K.A.; Gunning, P.T.; Glenn, M.; Katt, W.P.; Zhang, S.; Schroeck, C.; Sebti, S.M.; Jove, R.; Hamilton, A.D.; Turkson, J. An oxazole-based small-molecule Stat3 inhibitor modulates Stat3 stability and processing and induces antitumor cell effects. *ACS Chem. Biol.* **2007**, *2*, 787–798. [CrossRef]
88. Schust, J.; Sperl, B.; Hollis, A.; Mayer, T.U.; Berg, T. Stattic: A small-molecule inhibitor of STAT3 activation and dimerization. *Chem. Biol.* **2006**, *13*, 1235–1242. [CrossRef]
89. Matsuno, K.; Masuda, Y.; Uehara, Y.; Sato, H.; Muroya, A.; Takahashi, O.; Yokotagawa, T.; Furuya, T.; Okawara, T.; Otsuka, M.; et al. Identification of a New Series of STAT3 Inhibitors by Virtual Screening. *ACS Med. Chem. Lett.* **2010**, *1*, 371–375. [CrossRef] [PubMed]
90. Chen, H.; Yang, Z.; Ding, C.; Chu, L.; Zhang, Y.; Terry, K.; Liu, H.; Shen, Q.; Zhou, J. Fragment-based drug design and identification of HJC0123, a novel orally bioavailable STAT3 inhibitor for cancer therapy. *Eur. J. Med. Chem.* **2013**, *62*, 498–507. [CrossRef] [PubMed]
91. Kim, M.J.; Nam, H.J.; Kim, H.P.; Han, S.W.; Im, S.A.; Kim, T.Y.; Oh, D.Y.; Bang, Y.J. OPB-31121, a novel small molecular inhibitor, disrupts the JAK2/STAT3 pathway and exhibits an antitumor activity in gastric cancer cells. *Cancer Lett.* **2013**, *335*, 145–152. [CrossRef] [PubMed]
92. Yang, C.L.; Liu, Y.Y.; Ma, Y.G.; Xue, Y.X.; Liu, D.G.; Ren, Y.; Liu, X.B.; Li, Y.; Li, Z. Curcumin blocks small cell lung cancer cells migration, invasion, angiogenesis, cell cycle and neoplasia through Janus kinase-STAT3 signalling pathway. *PLoS ONE* **2012**, *7*, e37960. [CrossRef]
93. Fossey, S.L.; Bear, M.D.; Lin, J.; Li, C.; Schwartz, E.B.; Li, P.-K.; Fuchs, J.R.; Fenger, J.; Kisseberth, W.C.; London, C.A. The novel curcumin analog FLLL32 decreases STAT3 DNA binding activity and expression, and induces apoptosis in osteosarcoma cell lines. *BMC Cancer* **2011**, *11*, 1–15. [CrossRef]
94. Selvendiran, K.; Ahmed, S.; Dayton, A.; Kuppusamy, M.L.; Rivera, B.K.; Kálai, T.; Hideg, K.; Kuppusamy, P. HO-3867, a curcumin analog, sensitizes cisplatin-resistant ovarian carcinoma, leading to therapeutic synergy through STAT3 inhibition. *Cancer Biol. Ther.* **2011**, *12*, 837–845. [CrossRef]
95. Tierney, B.J.; McCann, G.A.; Cohn, D.E.; Eisenhauer, E.; Sudhakar, M.; Kuppusamy, P.; Hideg, K.; Selvendiran, K. HO-3867, a STAT3 inhibitor induces apoptosis by inactivation of STAT3 activity in BRCA1-mutated ovarian cancer cells. *Cancer Biol. Ther.* **2012**, *13*, 766–775. [CrossRef]
96. Onimoe, G.I.; Liu, A.; Lin, L.; Wei, C.C.; Schwartz, E.B.; Bhasin, D.; Li, C.; Fuchs, J.R.; Li, P.K.; Houghton, P.; et al. Small molecules, LLL12 and FLLL32, inhibit STAT3 and exhibit potent growth suppressive activity in osteosarcoma cells and tumor growth in mice. *Invest. New Drugs* **2012**, *30*, 916–926. [CrossRef] [PubMed]
97. Bid, H.K.; Oswald, D.; Li, C.; London, C.A.; Lin, J.; Houghton, P.J. Anti-angiogenic activity of a small molecule STAT3 inhibitor LLL12. *PLoS ONE* **2012**, *7*, e35513. [CrossRef]
98. Shakibaei, M.; Harikumar, K.B.; Aggarwal, B.B. Resveratrol addiction: To die or not to die. *Mol. Nutr. Food Res.* **2009**, *53*, 115–128. [CrossRef] [PubMed]
99. Kim, J.E.; Kim, H.S.; Shin, Y.-J.; Lee, C.S.; Won, C.; Lee, S.-A.; Lee, J.W.; Kim, Y.; Kang, J.-S.; Ye, S.-K. LYR71, a derivative of trimeric resveratrol, inhibits tumorigenesis by blocking STAT3-mediated matrix metalloproteinase 9 expression. *Exp. Mol. Med.* **2008**, *40*, 514–522. [CrossRef]
100. Sarkar, S.; Dutta, D.; Samanta, S.K.; Bhattacharya, K.; Pal, B.C.; Li, J.; Datta, K.; Mandal, C.; Mandal, C. Oxidative inhibition of Hsp90 disrupts the super-chaperone complex and attenuates pancreatic adenocarcinoma in vitro and *in vivo*. *Int. J. Cancer* **2013**, *132*, 695–706. [CrossRef] [PubMed]
101. Ito, C.; Itoigawa, M.; Nakao, K.; Murata, T.; Tsuboi, M.; Kaneda, N.; Furukawa, H. Induction of apoptosis by carbazole alkaloids isolated from Murraya koenigii. *Phytomedicine* **2006**, *13*, 359–365. [CrossRef]
102. Roy, M.K.; Thalang, V.N.; Trakoontivakorn, G.; Nakahara, K. Mahanine, a carbazole alkaloid from Micromelum minutum, inhibits cell growth and induces apoptosis in U937 cells through a mitochondrial dependent pathway. *Br. J. Pharmacol.* **2005**, *145*, 145–155. [CrossRef] [PubMed]
103. Roy, M.K.; Thalang, V.N.; Trakoontivakorn, G.; Nakahara, K. Mechanism of mahanine-induced apoptosis in human leukemia cells (HL-60). *Biochem. Pharmacol.* **2004**, *67*, 41–51. [CrossRef]
104. Sinha, S.; Pal, B.C.; Jagadeesh, S.; Banerjee, P.P.; Bandyopadhaya, A.; Bhattacharya, S. Mahanine inhibits growth and induces apoptosis in prostate cancer cells through the deactivation of Akt and activation of caspases. *Prostate* **2006**, *66*, 1257–1265. [CrossRef]
105. Sreedhar, A.S.; Söti, C.; Csermely, P. Inhibition of Hsp90: A new strategy for inhibiting protein kinases. *Biochim. Biophys. Acta Proteins Proteom.* **2004**, *1697*, 233–242. [CrossRef]

106. Saturnino, C.; Iacopetta, D.; Sinicropi, M.S.; Rosano, C.; Caruso, A.; Caporale, A.; Marra, N.; Marengo, B.; Pronzato, M.A.; Parisi, O.I. N-alkyl carbazole derivatives as new tools for Alzheimer's disease: Preliminary studies. *Molecules* **2014**, *19*, 9307–9317. [CrossRef]

107. Hou, S.; Yi, Y.W.; Kang, H.J.; Zhang, L.; Kim, H.J.; Kong, Y.; Liu, Y.; Wang, K.; Kong, H.-S.; Grindrod, S. Novel carbazole inhibits phospho-STAT3 through induction of protein–tyrosine phosphatase PTPN6. *J. Med. Chem.* **2014**, *57*, 6342–6353. [CrossRef]

108. Kong, Y.; Jung, M.; Wang, K.; Grindrod, S.; Velena, A.; Lee, S.A.; Dakshanamurthy, S.; Yang, Y.; Miessau, M.; Zheng, C. Histone deacetylase cytoplasmic trapping by a novel fluorescent HDAC inhibitor. *Mol. Cancer Ther.* **2011**, *10*, 1591–1599. [CrossRef]

109. Kong, H.-S.; Tian, S.; Kong, Y.; Du, G.; Zhang, L.; Jung, M.; Dritschilo, A.; Brown, M.L. Preclinical studies of YK-4-272, an inhibitor of class II histone deacetylases by disruption of nucleocytoplasmic shuttling. *Pharm. Res.* **2012**, *29*, 3373–3383. [CrossRef] [PubMed]

110. Walls, T.H.; Grindrod, S.C.; Beraud, D.; Zhang, L.; Baheti, A.R.; Dakshanamurthy, S.; Patel, M.K.; Brown, M.L.; MacArthur, L.H. Synthesis and biological evaluation of a fluorescent analog of phenytoin as a potential inhibitor of neuropathic pain and imaging agent. *Bioorg. Med. Chem.* **2012**, *20*, 5269–5276. [CrossRef]

111. Kang, H.J.; Yi, Y.W.; Hou, S.-J.; Kim, H.J.; Kong, Y.; Bae, I.; Brown, M.L. Disruption of STAT3-DNMT1 interaction by SH-I-14 induces re-expression of tumor suppressor genes and inhibits growth of triple-negative breast tumor. *Oncotarget* **2017**, *8*, 83457. [CrossRef]

112. Zhang, Q.; Wang, H.Y.; Marzec, M.; Raghunath, P.N.; Nagasawa, T.; Wasik, M.A. STAT3-and DNA methyltransferase 1-mediated epigenetic silencing of SHP-1 tyrosine phosphatase tumor suppressor gene in malignant T lymphocytes. *Proc. Natl. Acad. Sci. USA* **2005**, *102*, 6948–6953. [CrossRef]

113. Cuenca-López, M.D.; Serrano-Heras, G.; Montero, J.C.; Corrales-Sánchez, V.; Gomez-Juarez, M.; Gascón-Escribano, M.J.; Morales, J.C.; Voisin, V.; Núñez, L.E.; Morís, F. Antitumor activity of the novel multi-kinase inhibitor EC-70124 in triple negative breast cancer. *Oncotarget* **2015**, *6*, 27923. [CrossRef] [PubMed]

114. Estupiñan, O.; Santos, L.; Rodriguez, A.; Fernandez-Nevado, L.; Costales, P.; Perez-Escuredo, J.; Hermosilla, M.A.; Oro, P.; Rey, V.; Tornin, J. The multikinase inhibitor EC-70124 synergistically increased the antitumor activity of doxorubicin in sarcomas. *Int. J. Cancer* **2019**, *145*, 254–266. [CrossRef]

115. Botta, A.; Sirignano, E.; Popolo, A.; Saturnino, C.; Terracciano, S.; Foglia, A.; Sinicropi, M.S.; Longo, P.; Di Micco, S. Identification of lead compounds as inhibitors of STAT3: Design, synthesis and bioactivity. *Mol. Inform.* **2015**, *34*, 689–697. [CrossRef]

116. Zimmermann, K.; Sang, X.; Mastalerz, H.A.; Johnson, W.L.; Zhang, G.; Liu, Q.; Batt, D.; Lombardo, L.J.; Vyas, D.; Trainor, G.L. 9H-Carbazole-1-carboxamides as potent and selective JAK2 inhibitors. *Bioorg. Med. Chem. Lett.* **2015**, *25*, 2809–2812. [CrossRef]

117. Baburajeev, C.; Mohan, C.D.; Patil, G.S.; Rangappa, S.; Pandey, V.; Sebastian, A.; Fuchs, J.E.; Bender, A.; Lobie, P.E.; Rangappa, K.S. Nano-cuprous oxide catalyzed one-pot synthesis of a carbazole-based STAT3 inhibitor: A facile approach via intramolecular C–N bond formation reactions. *RSC Adv.* **2016**, *6*, 36775–36785. [CrossRef]

118. Diaz, T.; Navarro, A.; Ferrer, G.; Gel, B.; Gaya, A.; Artells, R.; Bellosillo, B.; Garcia-Garcia, M.; Serrano, S.; Martínez, A. Lestaurtinib inhibition of the Jak/STAT signaling pathway in hodgkin lymphoma inhibits proliferation and induces apoptosis. *PLoS ONE* **2011**, *6*, e18856. [CrossRef]

119. Shabbir, M.; Stuart, R. Lestaurtinib, a multitargeted tyrosinse kinase inhibitor: From bench to bedside. *Expert Opin. Investig. Drugs* **2010**, *19*, 427–436. [CrossRef]

120. Joos, S.; Küpper, M.; Ohl, S.; Von Bonin, F.; Mechtersheimer, G.; Bentz, M.; Marynen, P.; Möller, P.; Pfreundschuh, M.; Trümper, L. Genomic imbalances including amplification of the tyrosine kinase gene JAK2 in CD30+ Hodgkin cells. *Cancer Res.* **2000**, *60*, 549–552. [PubMed]

121. Amarante-Mendes, G.P.; McGahon, A.J.; Nishioka, W.K.; Afar, D.E.; Witte, O.N.; Green, D.R. Bcl-2-independent Bcr–Abl-mediated resistance to apoptosis: Protection is correlated with up regulation of Bcl-x L. *Oncogene* **1998**, *16*, 1383–1390. [CrossRef] [PubMed]

122. Weniger, M.; Melzner, I.; Menz, C.; Wegener, S.; Bucur, A.; Dorsch, K.; Mattfeldt, T.; Barth, T.; Möller, P. Mutations of the tumor suppressor gene SOCS-1 in classical Hodgkin lymphoma are frequent and associated with nuclear phospho-STAT5 accumulation. *Oncogene* **2006**, *25*, 2679–2684. [CrossRef]

123. Kube, D.; Holtick, U.; Vockerodt, M.; Ahmadi, T.; Haier, B.; Behrmann, I.; Heinrich, P.C.; Diehl, V.; Tesch, H. STAT3 is constitutively activated in Hodgkin cell lines. *Blood* **2001**, *98*, 762–770. [CrossRef] [PubMed]

124. Geyer, H.L.; Tibes, R.; Mesa, R.A. JAK2 inhibitors and their impact in myeloproliferative neoplasms. *Hematology* **2012**, *17*, s129–s132. [CrossRef]

125. Santos, F.P.; Kantarjian, H.M.; Jain, N.; Manshouri, T.; Thomas, D.A.; Garcia-Manero, G.; Kennedy, D.; Estrov, Z.; Cortes, J.; Verstovsek, S. Phase 2 study of CEP-701, an orally available JAK2 inhibitor, in patients with primary or post-polycythemia vera/essential thrombocythemia myelofibrosis. *Blood* **2010**, *115*, 1131–1136. [CrossRef]

126. Issa, S.; Prandina, A.; Bedel, N.; Rongved, P.; Yous, S.; Le Brogne, M.; Bouaziz, Z. Carbazole scaffolds in cancer therapy: A review from 2012 to 2018. *J. Enzym. Inhib Med. Chem* **2019**, *34*, 1321–1346. [CrossRef]

127. Tibes, R.; Bogenberger, J.M.; Benson, K.L.; Mesa, R.A. Current outlook on molecular pathogenesis and treatment of myeloproliferative neoplasm. *Mol. Diagn. Ther.* **2012**, *16*, 269–283. [CrossRef] [PubMed]

*Article*

# A Nitrocarbazole as a New Microtubule-Targeting Agent in Breast Cancer Treatment

Maria Stefania Sinicropi [1,†], Cinzia Tavani [2,†], Camillo Rosano [3], Jessica Ceramella [1,*], Domenico Iacopetta [1], Alexia Barbarossa [4], Lara Bianchi [2], Alice Benzi [2], Massimo Maccagno [2], Marco Ponassi [3], Domenico Spinelli [5] and Giovanni Petrillo [2,*]

[1] Department of Pharmacy, Health and Nutritional Sciences, University of Calabria, 87036 Rende, Italy; s.sinicropi@unical.it (M.S.S.); domenico.iacopetta@unical.it (D.I.)

[2] Department of Chemistry and Industrial Chemistry, University of Genoa, Via Dodecaneso 31, 16146 Genoa, Italy; cinzia.tavani@unige.it (C.T.); lara.bianchi@unige.it (L.B.); alice.benzi@edu.unige.it (A.B.); massimo.maccagno@unige.it (M.M.)

[3] Proteomics and Mass Spectrometry Unit, IRCCS Policlinico San Martino, Largo Rosanna Benzi, 10, 16132 Genoa, Italy; camillo.rosano@hsanmartino.it (C.R.); marco.ponassi@hsanmartino.it (M.P.)

[4] Department of Pharmacy-Drug Sciences, University of Bari "Aldo Moro", 70126 Bari, Italy; alexia.barbarossa@uniba.it

[5] Department of Chemistry 'G. Ciamician', Alma Mater Studiorum, University of Bologna, Via Selmi 2, 40126 Bologna, Italy; domenico.spinelli@unibo.it

[*] Correspondence: jessicaceramella@gmail.com (J.C.); giovanni.petrillo@unige.it (G.P.); Tel.: +39-098-449-3200 (J.C.); +39-010-353-6103 (G.P.)

[†] These authors equally contributed to this work.

**Citation:** Sinicropi, M.S.; Tavani, C.; Rosano, C.; Ceramella, J.; Iacopetta, D.; Barbarossa, A.; Bianchi, L.; Benzi, A.; Maccagno, M.; Ponassi, M.; et al. A Nitrocarbazole as a New Microtubule-Targeting Agent in Breast Cancer Treatment. *Appl. Sci.* **2021**, *11*, 9139. https://doi.org/10.3390/app11199139

Academic Editor: Caterina Ledda

Received: 27 July 2021
Accepted: 23 September 2021
Published: 1 October 2021



**Abstract:** Breast cancer is still considered a high-incidence disease, and numerous are the research efforts for the development of new useful and effective therapies. Among anticancer drugs, carbazole compounds are largely studied for their anticancer properties and their ability to interfere with specific targets, such as microtubule components. The latter are involved in vital cellular functions, and the perturbation of their dynamics leads to cell cycle arrest and subsequent apoptosis. In this context, we report the anticancer activity of a series of carbazole analogues **1–8**. Among them, 2-nitrocarbazole **1** exhibited the best cytotoxic profile, showing good anticancer activity against two breast cancer cell lines, namely MCF-7 and MDA-MB-231, with $IC_{50}$ values of $7 \pm 1.0$ and $11.6 \pm 0.8$ μM, respectively. Furthermore, compound **1** did not interfere with the growth of the normal cell line MCF-10A, contrarily to Ellipticine, a well-known carbazole derivative used as a reference molecule. Finally, in vitro immunofluorescence analysis and in silico studies allowed us to demonstrate the ability of compound **1** to interfere with tubulin organization, similarly to vinblastine: a feature that results in triggering MCF-7 cell death by apoptosis, as demonstrated using a TUNEL assay.

**Keywords:** carbazoles; ellipticine; tubulin; breast cancer; apoptosis; docking simulations

## 1. Introduction

Despite many research efforts, breast cancer incidence is constantly growing and remains a serious health emergency [1]. Indeed, breast cancer represents, to date, the main cancer-related cause of disease for women, and its diagnosis and mortality frequencies have risen worldwide in recent years [2]. Among the estimated 19.3 million new cancer cases worldwide in 2020, 11.7% represents female breast cancer, correlated with a mortality of 6.9% [3].

Clinically, breast cancers are classified according to specific subtypes defined by their histopathological features, and their expression of hormone receptors and growth factors, such as estrogen receptor (ER), progesterone receptor (PR), and human epidermal growth factor receptor 2 (HER2). ER-positive breast cancer is increasing in incidence, while the triple negative causes concern about its aggressiveness and ability to give rise to metastases [4,5].

The use of cytotoxic chemotherapy in breast cancer has made significant progress in recent years due to the use of drugs able to interfere with the numerous biological pathways involved in cancer cell growth. However, the numerous side effects related to current therapies often overshadow their benefits. This spurred the need for research and development of new potent anticancer agents. Currently, medical attention has primarily focused on naturally occurring molecules with anticancer properties. Among them, the carbazole scaffold represents an important structural motif of many natural and/or synthetic pharmacologically active compounds [6]. They have been found in a large variety of organisms, including bacteria, fungi, plants, and animals and represent an important class of heterocycles, which exhibits innumerable biological activities [7]. Ellipticine is an alkaloid considered the first lead compound with anticancer activity belonging to the carbazole class [8]. It was first obtained in 1959 from the leaves of *Ochrosia elliptica* (Apocynacae), while now, it is prepared by entirely synthetic procedures [9]. Considering the biological importance of this molecule together with its demonstrated high toxicity [10], many Ellipticine derivatives with antioxidant, anticancer, anti-inflammatory, antibacterial, antiviral, and antidiabetic properties have been synthesized in recent decades [7,11–16]. Numerous in vitro studies, supported by docking simulations, demonstrated that some carbazole derivatives and analogues significantly disrupt the microtubule network, arresting the cell cycle and inducing cell apoptosis [17–20].

Considering these exciting data [17–20], the goal of this work was to evaluate the anticancer activity of a series of carbazole derivatives (**1–8**, Figure 1) against two human breast cancer cell lines, namely ER(+) MCF-7 cells and triple-negative MDA-MB-231 cells. The obtained data showed that nitrocarbazoles **1–3** exhibited the best anticancer activity on both the breast cancer cell lines used. However, 3-nitrocarbazole **2** and 2,3-dinitrocarbazole **3** showed strong cytotoxicity on the normal MCF-10A cell line, while 2-nitrocarbazole **1** did not interfere with the growth of the same cell line. Cytotoxicity of compounds on normal cells may be influenced by the position of the nitro group(s) on the aromatic ring. These interesting results pushed us to further understand the mechanism of action of the most active and safe nitrocarbazole **1** in depth. Immunofluorescence analysis demonstrated that compound **1** perturbs microtubule networks, inducing disorganization of the tubulin filaments and their accumulation around cell nuclei. The disruption of microtubules dynamics led to cancer cell death by apoptosis. The in vitro results, confirmed by in silico studies, suggest that 2-nitrocarbazole **1** represents an interesting tool in cancer treatment as a microtubule-targeting agent. These results are an important starting point in medicinal chemistry for the development of targeted therapy able to reduce the numerous toxic effects typically associated with traditional therapeutic approaches.

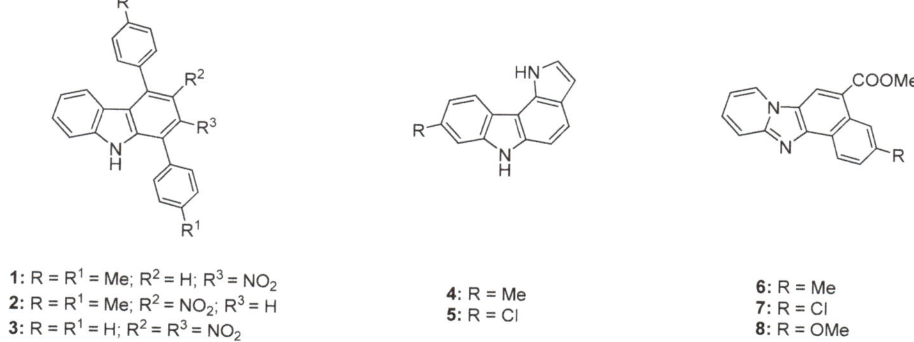

**1:** R = R$^1$ = Me; R$^2$ = H; R$^3$ = NO$_2$
**2:** R = R$^1$ = Me; R$^2$ = NO$_2$; R$^3$ = H
**3:** R = R$^1$ = H; R$^2$ = R$^3$ = NO$_2$

**4:** R = Me
**5:** R = Cl

**6:** R = Me
**7:** R = Cl
**8:** R = OMe

**Figure 1.** Molecular structures of compounds **1–8**.

## 2. Materials and Methods

### 2.1. Chemistry

The synthesis and characterization of compounds **1–3** [21], **4** and **5** [22], and **6–8** [23] has already been reported; details are provided in Scheme 1 below.

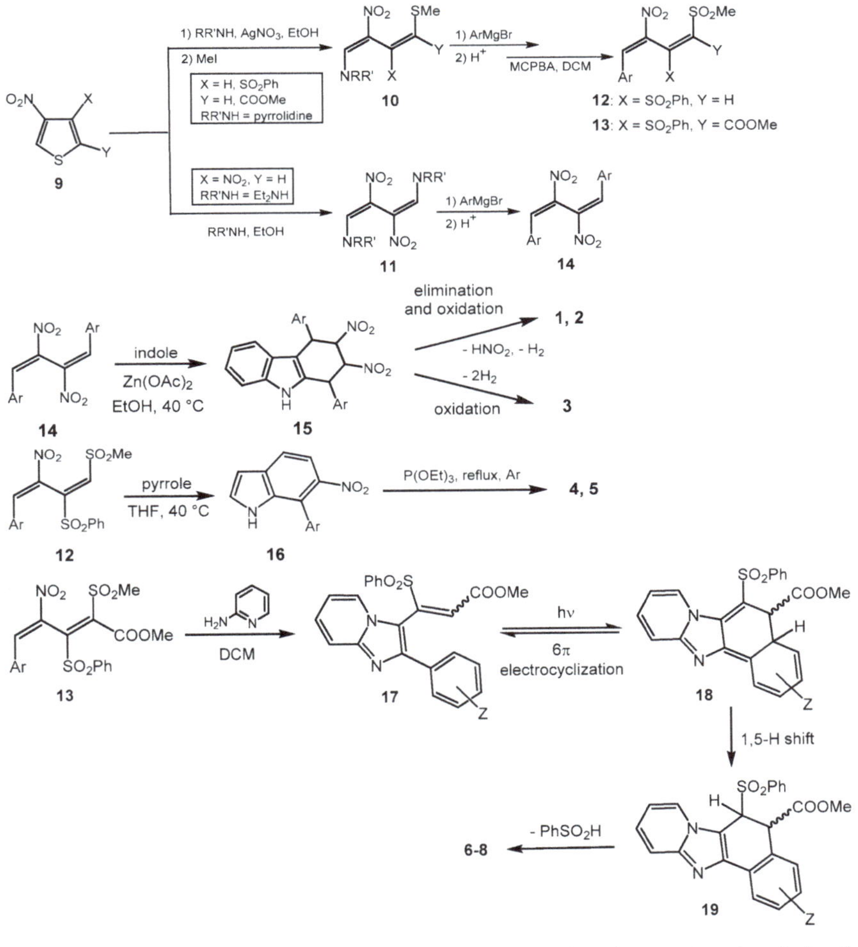

**Scheme 1.** Synthetic protocol of compounds **1–8**. From nitrothiophene **9**, via a ring-opening–ring-closing (benzannulation) procedure.

### 2.2. Biology

#### 2.2.1. Cell Culture

The three cell lines employed in this work (MCF-7, MDA-MB-231, and MCF-10A) were purchased from American Type Culture Collection (ATCC, Manassas, VA, USA) and cultured as already described [15].

#### 2.2.2. MTT Assay

The in vitro anticancer activity of all of the studied compounds were detected using the MTT (Sigma) assay [24,25]. In brief, cells were seeded in a 48-well plate, then starved in serum-free medium, and incubated with the target compounds dissolved in

DMSO at six differing concentrations (0.1, 1, 5, 10, 100, and 200 µM) for 72 h, as already described [24]. After this period of time, MTT (3-(4,5-dimethylthiazol-2-yl)-2,5-diphenyl tetrazolium bromide) was incubated for 2 h at 37 °C (final concentration 0.5 mg/mL). Then, the formazan crystals were dissolved in DMSO and the optical density was measured at 570 nm using a microplate reader. All of the calculations were performed in triplicate, and the results were represented as the percent (%) of basal. The $IC_{50}$ values were calculated using curve-fitting GraphPad Prism 9 software (GraphPad Software, La Jolla, CA, USA) with nonlinear regression. The values represent the mean ± standard deviation ($n = 3$).

### 2.2.3. Immunofluorescence Analysis

The cells were seeded in 48-well culture plates containing glass slides, then serum-deprived for 24 h, and incubated with the most active compound for 24 h (concentration equal to its $IC_{50}$ value). Then, the methanol-fixed cells were incubated with the primary antibody (mouse anti-β-tubulin, Santa Cruz Biotechnology, Dallas, TX, USA) and then the secondary antibody (Alexa Fluor® 568 conjugate goat-anti-mouse, Thermo Fisher Scientific, MA, USA), as previously described [26]. Then, the Nuclei were stained using DAPI (Sigma Aldrich, Milan, Italy). Fluorescence was detected using a fluorescence microscope (Leica DM 6000, 20× magnification). LAS-X software was used to acquire and process all images. The images are representative of three independent experiments.

### 2.2.4. Docking Studies

The crystal structures of the quaternary assembly of human tubulin (αβαβ) in a complex with stathmin and vinblastine [27] (PDB code 5J2T) has been used as a target for our molecular docking simulations. We built the three-dimensional structures of compounds **1**, **2**, and **3** using the MarvinSketch program (ChemAxon ltd, Budapest, Hu), and once the atomic charges were assigned, we minimized all of them. As described in our previous work [15], we used the Autodock program v. 4.2.2 [28] to evaluate the possible binding modes of our ligands and to evaluate the binding energies of different derivatives to these proteins. We adopted a "blind docking" strategy: docking simulations of small molecules to the targets were conducted without a priori knowledge of the position of the binding site by the system. All of the simulations were performed by adopting the standard default values and by utilizing the same procedures described in several previous work by our group [29–31]. The figures were drawn using the program Chimera [32].

### 2.2.5. TUNEL Assay

The ability of the most active compound to induce cell death by apoptosis was detected using the TUNEL assay using the CF™488A TUNEL Assay Apoptosis Detection Kit (Biotium, Hayward, CA, USA). The cells were grown on glass coverslips and then treated with the tested compound. Then, the methanol-fixed cells were incubated with the enzyme terminal deoxynucleotidyl transferase (TdT) for 2 h at 37 °C, as previously described [24]. The nuclei were stained using DAPI 0.2 mg/mL (Sigma Aldrich, Milan, Italy). Finally, the cells were observed under a fluorescence microscope (Leica DM6000; 20x magnification). The images are demonstrative of three separate experiments.

## 3. Results and Discussion

### 3.1. Chemistry

Although structurally different, compounds **1–8** share a synthetic protocol whereby a nitrobutadiene (**12–14**), deriving from the initial ring-opening of a suitably substituted 3-nitrothiophene (**9**) [33–35], acts as a benzannulating agent towards indole [21], pyrrole [22], or 2-aminopyridine [23] (Scheme 1).

Evaluation of the anticancer activity of compounds **1–8** follows a long-standing engagement by the Genoa research group in the synthesis of pharmacologically active nitroderivatives from the initial ring-opening of nitrothiophenes **9** [36–44], an effort that has resulted in a number of positive results in both in vitro and in vivo experiments. For instance, appreciable antitumor activity has been found for either some modified **13** [36,38–42] or some selected **14** [36,37,39,40,42]; furthermore, α-glucosidase inhibition [43] or antibacterial activity [44] has, on the one hand, exhibited in nitroheterocycles obtained from our nitrobutadiene building-blocks. On the other hand, the Cosenza research group has, in turn, recently highlighted the efficacy of a nitrocarbazole as an anti-HIV agent [11]. Coupled with the abovementioned results, the outcomes herein surely contribute to assessing the nitro goup's significance as a valuable pharmacophore.

### 3.2. Biology

#### 3.2.1. Anticancer Activity

The anticancer activity of all the compounds against two breast cancer cell lines, namely ER(+) MCF-7 and triple-negative MDA-MB-231, were evaluated by MTT assay, and the $IC_{50}$ values, derived from the experimental data, are summarized in Table 1.

**Table 1.** $IC_{50}$ values of compounds **1–8** and Ellipticine, expressed in μM. The means ± standard deviations are shown. The experiments were performed in triplicate.

| Compound | MDA-MB-231 | MCF-7 | MCF-10A |
|---|---|---|---|
| 1 | 11.6 ± 0.8 | 7.0 ± 1.0 | >200 |
| 2 | 12.2 ± 1.2 | 3.4 ± 1.3 | 23.6 ± 0.7 |
| 3 | 14.4 ± 0.9 | 5.4 ± 1.1 | 3.7 ± 0.6 |
| 4 | >200 | 162.5 ± 1.4 | 110.5 ± 0.9 |
| 5 | 1.2 ± 1.1 | 1.7 ± 0.6 | 27.8 ± 1.0 |
| 6 | >200 | >200 | >200 |
| 7 | >200 | >200 | 143.3 ± 1.1 |
| 8 | >200 | >200 | >200 |
| Ellipticine | 1.3 ± 0.9 | 1.9 ± 0.5 | 1.2 ± 0.7 |

After the breast cancer cells were incubated in the presence of compounds **1–8** for 72 h, the $IC_{50}$ values indicated that some of them exhibited, in different degrees, a good anticancer activity against both cell lines.

The most promising compound was 2-nitrocarbazole **1**, which exerted good anticancer activity against both breast cancer cell lines used in this assay, with $IC_{50}$ values of 7.0 ± 1.0 and 11.6 ± 0.8 μM on MCF-7 and MDA-MB-231, respectively. Carbazoles **2**, **3**, and **5** showed much higher cytotoxicity than the other compounds on both of the cell lines screened. Indeed, the $IC_{50}$ values of compounds **2**, **3**, and **5** were 3.4 ± 1.3, 5.4 ± 1.1, and 1.7 ± 0.6 μM against MCF-7 cells, respectively, and 12.2 ± 1.2, 14.4 ± 0.9, and 1.2 ± 1.1 μM towards MDA-MB-231 cells, respectively. Unfortunately, together with their good anticancer activity, **2**, **3**, and **5** also exhibited severe cytotoxicity on the normal human mammary epithelial cells MCF-10A, with $IC_{50}$ values of 23.6 ± 0.7, 3.7 ± 0.6, and 27.8 ± 1.0 μM, respectively.

Instead, **1** did not interfere with the growth of the normal cell lines, showing an $IC_{50}$ value higher than 200 μM on the same normal cells. Moreover, compounds **4** and **6–8** exhibited a lower, or no, anticancer activity against both breast cancer cell lines. Ellipticine, used as a reference molecule in this assay, exhibited strong anticancer activity against MCF-7 and MDA-MB-231 cells with $IC_{50}$ values of 1.9 ± 0.5 and 1.3 ± 0.9 μM, respectively, together with a dramatic inhibition of normal cell growth (the $IC_{50}$ on MCF-10A was 1.2 ± 0.7 μM). Concerning the structure–activity relationships, the nitrocarbazoles **1–3** resulted in the most active compounds, indicating that the presence of the nitro group could positively affect their anticancer activity. However, it seems clear that the position of the nitro group on the aromatic ring influences the cytotoxicity of the compounds on the normal cells. Indeed, if present in the 2-position (2-nitrocarbazole **1**), it does not affect the growth of normal cells MCF-10A ($IC_{50}$ > 200 μM), at least at the concentrations and under the conditions used, while in the 3-position, it makes compound **2** responsible for the strong cytotoxic effect on the same non-tumoral cells ($IC_{50}$ = 23.6 ± 0.7 μM). Moreover, the presence of two nitro groups in the 2,3-positions on the aromatic ring gives a higher toxicity to compound **3** ($IC_{50}$ = 3.7 ± 0.6 µM) than to **2**.

Regarding the 8-methylpyrrolo[3,2-*c*]carbazole **4**, it did not show an anticancer effect on the breast cancer cells used, whereas the replacement of the methyl group with a chlorine makes the 8-chloropyrrolo[3,2-*c*]carbazole **5** more active ($IC_{50}$ = 1.7 ± 0.6 µM on MCF-7 and 1.2 ± 1.1 µM on MDA-MB-231) and highly cytotoxic on the normal cells MCF-10A ($IC_{50}$ = 27.8 ± 1.0 µM). Finally, imidazopyridine **6–8** were inactive as anticancer agents on both breast cancer cell lines.

Summing up, among the tested compounds, the 2-nitrocarbazole **1** possesses the best anticancer profile, causing a growth reduction in both breast cancer cells used, even if less active than the reference molecule Ellipticine. However, contrary to the other analogues of the series and to Ellipticine, the 2-nitrocarbazole **1** did not exert any cytotoxicity against normal human mammary epithelial cells MCF-10A.

### 3.2.2. Induction of Cell Cytoskeleton Destabilization

Microtubule-targeting agents are widespread drugs useful in cancer treatment due to their ability to interfere with critical cellular functions, such as mitosis, cell migration, and cell signaling [45].

The efficacy of microtubule-targeting drugs has been evidenced by the use of some Vinca alkaloids and taxanes for the treatment of a large panel of human cancers [46]. Based on the mechanism of action, microtubule-targeting agents are classified into two categories: microtubule-destabilizing agents, such as the Vinca alkaloids, that inhibit the polymerization reaction; destabilizing microtubules and decreasing tubulin polymer filaments; and microtubule-stabilizing agents, such as taxanes which, contrarily, promote tubulin polymerization-stabilizing microtubules [47]. While these agents are highly effective in cancer treatment, the onset of resistance represents the principal clinical issue that limits their use [48]. Moreover, their effectiveness has been impaired by the presence of systemic toxicity and, often, the absence of bioavailability [49]. Thus, in recent years, research efforts have been focused on the development of more active and safe new compounds that could target microtubule organization [50–52].

With the aim to understand the role of compound **1** in cytoskeleton dynamics, we carried out immunofluorescence studies using MCF-7 cells as models, since they represent the cell line on which the 2-nitrocarbazole **1** was more active. Cells treated only with the vehicle (DMSO and CTRL) showed normal organization of the microtubule network in which tubulin filaments are regularly spread into the MCF-7 cell cytoplasm (Figure 2, Panel B, CTRL). Contrarily, exposure of the same cells to vinblastine as well as to compound **1** caused microtubule disorganization (Figure 2, panels B, vinblastine and **1**). Indeed, tubulin filaments become irregular and accumulate around cell nuclei (see the white arrows). These results indicate that, similar to vinblastine, 2-nitrocarbazole **1** could act as a tubulin-polymerization inhibitor.

### 3.2.3. Docking Studies

To study the possible binding modes of our compounds to the quaternary assembly of human tubulin and to calculate a binding energy of the complexes, we carried over molecular docking simulation runs using the crystallographic coordinates of the complex formed between tubulin α, tubulin β, and stathmin4 (PDB code 5J2T) as a protein target [27], eventually comparing our results with the binding modes of vinblastine. For all our compounds, we adopted a "blind docking" strategy: i.e., the docking of our small molecules to their targets was performed without a priori knowledge of the binding site of the ligand. This strategy was first tested and validated by repositioning vinblastine in the protein binding site with its correct binding mode, displaying a root-mean-square deviation (RMSD) of less than 0.2 Å, compared with the one determined by X-ray crystallography. This guarantees the reliability of our docking simulations. Further on, we tested our compounds and found two different binding zones: one for molecules **1** and **3**, within the interface between subunits β and α in proximity to the vinblastine binding zone, and a second for **2**, at the interface between subunit α and β, in proximity to the Colchicine binding site (Figure 3). Table 2 illustrates the amino acids involved in ligand interactions and the binding energies of all of the complexes formed by tubulin and our compounds.

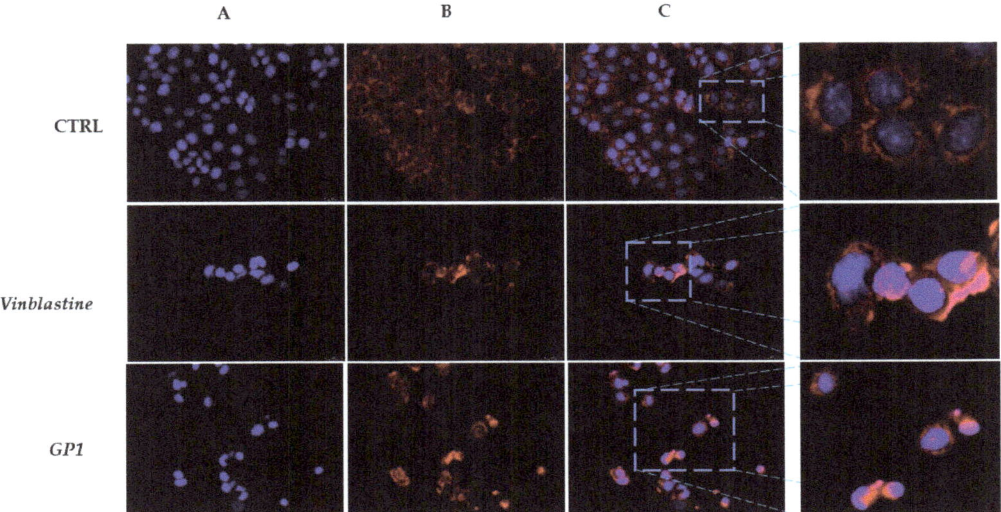

**Figure 2.** Immunofluorescence studies. MCF-7 cells were treated with compound **1** and vinblastine, used as a reference molecule, (IC$_{50}$ values) or with a vehicle (CTRL) for 24 h. After treatment, the cells were incubated with primary and secondary antibodies (see the Experimental section for more details) and then imaged under the inverted fluorescence microscope at 20× magnification. CTRL cells (DMSO only) showed regular organization of the cytoskeleton ((**B**), CTRL). MCF-7 cells treated with vinblastine and compound **1** exhibited irregular arrangement and organization, with tubulin filaments accumulated around cell nuclei (white arrows, (**B**), vinblastine and **1**). (**A**) DAPI, excitation/emission wavelength 350 nm/460 nm; (**B**) tubulin (Alexa Fluor® 568) excitation/emission wavelength 644 nm/665 nm; and (**C**) a merge. A zoom-in of the overlay channels is shown on the right. Images are representative of three independent experiments.

**Table 2.** Energies of the complexes formed by tubulin and compounds **1**, **2**, and **3**, and protein residues interacting with the ligands.

| Compound | Binding Energy (Kcal/mol) | Ki (nM) * | Atoms Involved in Binding § | | |
|---|---|---|---|---|---|
| | | | Protein residue | Distance (Å) | Residues involved in hydrophobic interactions |
| **1** | −9.76 | 70.4 | *Glnβ11* *Serβ40* *Thrβ145* | **3.01** **3.0** **2.42** | *Alaβ99, Leuβ141, Proβ173, Valβ177* |
| **2** | −10.65 | 15.56 | *Argα221* *Proα222* *Thrβ353* | **2.9** **3.2** **2.87** | *Valα172, Tyrα224, Leuβ248, Valβ335* |
| **3** | −9.49 | 110.58 | *Lysβ176* *Tyrβ219* *Thrβ221* | **2.76** **2.66** **2.92** | *Leuα248, Proα325, Alaα330, Tyrβ210, Proβ222, Tyrβ224* |

* K$_i$ values as calculated by the Autodock algorithm: Ki = exp(ΔG/(R*T). § Residues involved in hydrogen bonding are listed in **bold**. Hydrophobic contacts are listed in *italic*.

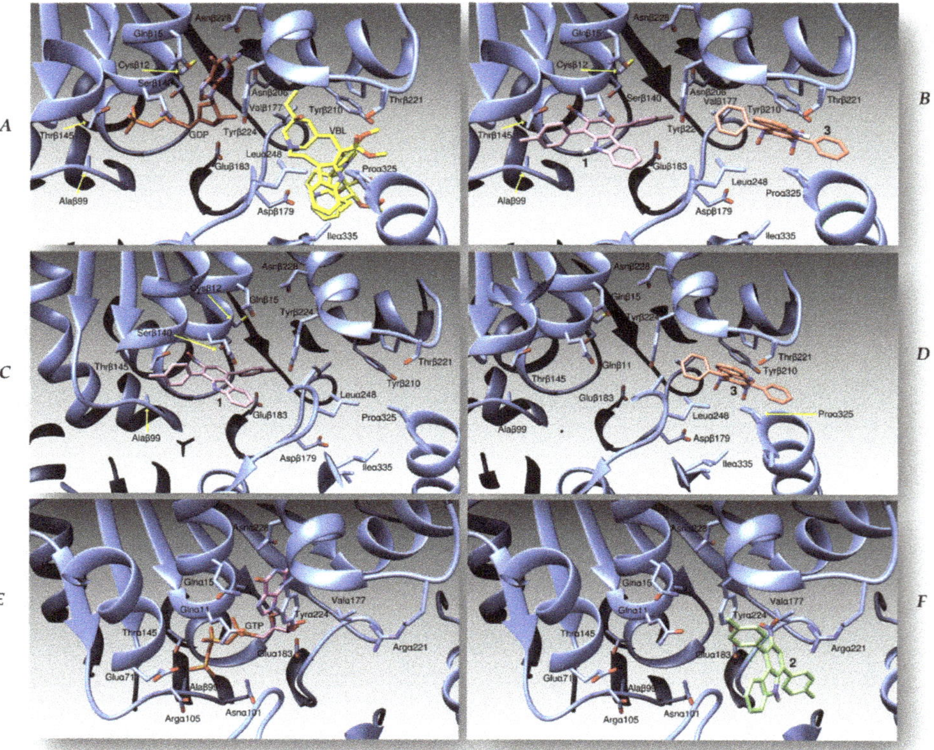

**Figure 3.** (**A**) A cartoon representation of the tubulin (cyan ribbons) interface between subunits beta and alpha, with vinblastine (VLB, yellow sticks) and guanosyndiphosphate (GDP, red sticks) reported. (**B**) Binding modes of molecules **1** (purple sticks) and **3** (pink) to the beta-alpha interface of tubulin, in the same three-dimensional orientation as (**A**). (**C,D**) The position of compounds **1** (purple) and **3** (pink), respectively, within the same binding site. (**E**) The crystallographic structure of a guanosyntriphosphate (GTP) molecule (drawn in purple) bound to the alpha–beta interface of the protein. (**F**) The pose of compound **2** (green sticks) within the same binding site. We showed and labelled the residues involved in the interactions of the three molecules in the tubulin quaternary structure.

### 3.2.4. TUNEL Assay

The perturbation of microtubule dynamics leads to disruption of the mitotic spindle in dividing cells, causing cell cycle arrest and, as a last step, the induction of subsequent cell death by apoptosis [53,54]. Thus, we determined whether 2-nitrocarbazole **1** is able to trigger apoptosis using TUNEL assay, which allows the formation of DNA fragments to be detected.

When compared with the vehicle-treated cells (CTRL), a detectable level of green fluorescence, related to the formation of DNA fragments, was evident in MCF-7 cells after 24 h treatment with compound **1** at its $IC_{50}$ value (Figure 4). This evidence indicates that this compound is able to induce MCF-7 cell death by triggering apoptosis, and this effect is probably linked to its ability to perturb microtubule dynamics.

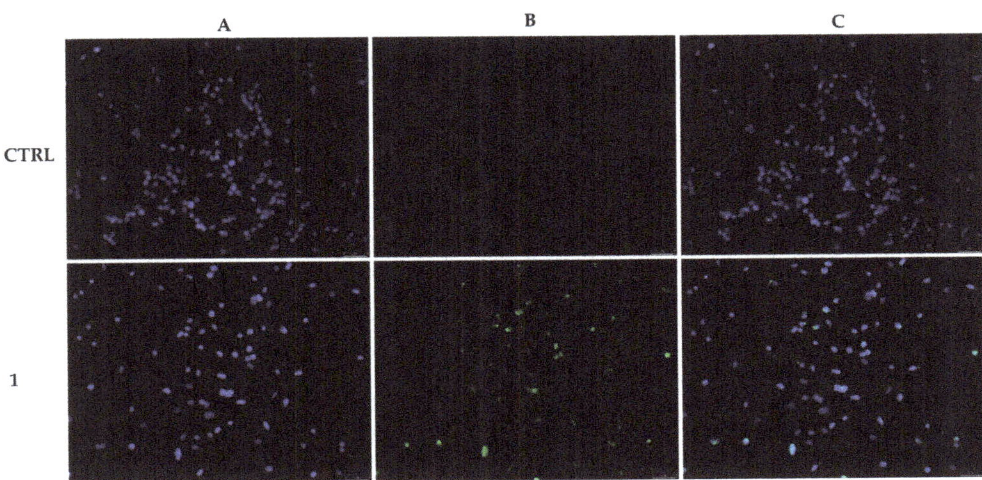

**Figure 4.** MCF-7 cells were exposed to compound **1** at its IC$_{50}$ value or with the vehicle DMSO (CTRL) for 24 h. Then, cells were incubated with the TdT enzyme and observed under an inverted fluorescence microscope at 20× magnification. Apoptotic cells are indicated by a clear green nuclear fluorescence in compound **1**-treated cells. (**A**) DAPI (CTRL and **1**) $\lambda_{ex/em}$ = 350 nm/460 nm. (**B**) CF$^{TM}$488A (CTRL and **1**) $\lambda_{ex/em}$ = 490 nm/515 nm. (**C**) A merge. Fields are representative of three separate experiments.

## 4. Conclusions

In this paper, we reported the evaluation of the anticancer properties of some nitro-carbazoles in in silico and in vitro studies. The lead compound showed a good cytotoxic profile, being active mostly against MCF-7 cells. Moreover, docking simulations and immunofluorescence studies suggest a role in perturbing the MCF-7 cell microtubule network and triggering cancer cell death by apoptosis.

**Author Contributions:** Conceptualization, G.P. and M.S.S.; methodology, D.I.; software, C.R. and M.P.; validation, D.S., formal analysis, J.C. and C.T.; investigation, J.C. and C.T.; data curation, A.B. (Alexia Barbarossa); writing—original draft preparation, J.C. and G.P.; writing—review and editing, M.S.S. and C.T.; visualization, A.B. (Alice Benzi), L.B., and M.M.; supervision, D.S. and G.P.; funding acquisition, C.T. and G.P. All authors have read and agreed to the published version of the manuscript.

**Funding:** Financial support was provided by grants from Genoa University (FRA 2020).

**Institutional Review Board Statement:** Not applicable.

**Informed Consent Statement:** Not applicable.

**Data Availability Statement:** Not applicable.

**Acknowledgments:** C.R. acknowledges the support of a grant from the Italian Ministry of Health (Ricerca Corrente 2020).

**Conflicts of Interest:** The authors declare no conflict of interest.

## References

1. Global Burden of Disease Cancer Collaboration; Fitzmaurice, C.; Akinyemiju, T.F.; Al Lami, F.H.; Alam, T.; Alizadeh-Navaei, R.; Allen, C.; Alsharif, U.; Alvis-Guzman, N.; Amini, E.; et al. Global, Regional, and National Cancer Incidence, Mortality, Years of Life Lost, Years Lived With Disability, and Disability-Adjusted Life-Years for 29 Cancer Groups, 1990 to 2016: A Systematic Analysis for the Global Burden of Disease Study. *JAMA Oncol.* **2018**, *4*, 1553–1568.
2. Ghoncheh, M.; Pournamdar, Z.; Salehiniya, H. Incidence and Mortality and Epidemiology of Breast Cancer in the World. *Asian Pac. J. Cancer Prev. APJCP* **2016**, *17*, 43–46. [CrossRef]

3. Sung, H.; Ferlay, J.; Siegel, R.L.; Laversanne, M.; Soerjomataram, I.; Jemal, A.; Bray, F. Global cancer statistics 2020: GLOBOCAN estimates of incidence and mortality worldwide for 36 cancers in 185 countries. *CA Cancer J. Clin.* **2021**, *71*, 209–249. [CrossRef]

4. Bigaard, J.; Stahlberg, C.; Jensen, M.B.; Ewertz, M.; Kroman, N. Breast cancer incidence by estrogen receptor status in Denmark from 1996 to 2007. *Breast Cancer Res. Treat.* **2012**, *136*, 559–564. [CrossRef]

5. Zhang, H.; Wang, J.; Yin, Y.; Meng, Q.; Lyu, Y. The role of EMT-related lncRNA in the process of triple-negative breast cancer metastasis. *Biosci. Rep.* **2021**, *41*, BSR20203121. [CrossRef]

6. Schmidt, A.W.; Reddy, K.R.; Knolker, H.J. Occurrence, biogenesis, and synthesis of biologically active carbazole alkaloids. *Chem. Rev.* **2012**, *112*, 3193–3328. [CrossRef]

7. Issa, S.; Prandina, A.; Bedel, N.; Rongved, P.; Yous, S.; Le Borgne, M.; Bouaziz, Z. Carbazole scaffolds in cancer therapy: A review from 2012 to 2018. *J. Enzym. Inhib. Med. Chem.* **2019**, *34*, 1321–1346. [CrossRef]

8. Stiborova, M.; Sejbal, J.; Borek-Dohalska, L.; Aimova, D.; Poljakova, J.; Forsterova, K.; Rupertova, M.; Wiesner, J.; Hudecek, J.; Wiessler, M.; et al. The anticancer drug ellipticine forms covalent DNA adducts, mediated by human cytochromes P450, through metabolism to 13-hydroxyellipticine and ellipticine N2-oxide. *Cancer Res.* **2004**, *64*, 8374–8380. [CrossRef] [PubMed]

9. Miller, C.M.; McCarthy, F.O. Isolation, biological activity and synthesis of the natural product ellipticine and related pyridocarbazoles. *RSC Adv.* **2012**, *2*, 8883–8918. [CrossRef]

10. Stiborova, M.; Poljakova, J.; Martinkova, E.; Borek-Dohalska, L.; Eckschlager, T.; Kizek, R.; Frei, E. Ellipticine cytotoxicity to cancer cell lines—A comparative study. *Interdiscip. Toxicol.* **2011**, *4*, 98–105. [CrossRef] [PubMed]

11. Caruso, A.; Ceramella, J.; Iacopetta, D.; Saturnino, C.; Mauro, M.V.; Bruno, R.; Aquaro, S.; Sinicropi, M.S. Carbazole Derivatives as Antiviral Agents: An Overview. *Molecules* **2019**, *24*, 1912. [CrossRef]

12. Caruso, A.; Lancelot, J.C.; El-Kashef, H.; Sinicropi, M.S.; Legay, R.; Lesnard, A.; Rault, S. A rapid and versatile synthesis of novel pyrimido[5,4-b]carbazoles. *Tetrahedron* **2009**, *65*, 10400–10405. [CrossRef]

13. Caruso, A.; Sinicropi, M.S.; Lancelot, J.C.; El-Kashef, H.; Saturnino, C.; Aubert, G.; Ballandonne, C.; Lesnard, A.; Cresteil, T.; Dallemagne, P.; et al. Synthesis and evaluation of cytotoxic activities of new guanidines derived from carbazoles. *Bioorg. Med. Chem. Lett.* **2014**, *24*, 467–472. [CrossRef] [PubMed]

14. Rizza, P.; Pellegrino, M.; Caruso, A.; Iacopetta, D.; Sinicropi, M.S.; Rault, S.; Lancelot, J.C.; El-Kashef, H.; Lesnard, A.; Rochais, C.; et al. 3-(Dipropylamino)-5-hydroxybenzofuro[2,3-f]quinazolin-1(2H)-one (DPA-HBFQ-1) plays an inhibitory role on breast cancer cell growth and progression. *Eur. J. Med. Chem.* **2016**, *107*, 275–287. [CrossRef] [PubMed]

15. Sinicropi, M.S.; Iacopetta, D.; Rosano, C.; Randino, R.; Caruso, A.; Saturnino, C.; Muia, N.; Ceramella, J.; Puoci, F.; Rodriquez, M.; et al. N-thioalkylcarbazoles derivatives as new anti-proliferative agents: Synthesis, characterisation and molecular mechanism evaluation. *J. Enzym. Inhib. Med. Chem.* **2018**, *33*, 434–444. [CrossRef] [PubMed]

16. Caruso, A.; Iacopetta, D.; Puoci, F.; Cappello, A.R.; Saturnino, C.; Sinicropi, M.S. Carbazole Derivatives: A Promising Scenario for Breast Cancer Treatment. *Mini-Rev. Med. Chem.* **2016**, *16*, 630–643. [CrossRef] [PubMed]

17. Ceramella, J.; Caruso, A.; Occhiuzzi, M.A.; Iacopetta, D.; Barbarossa, A.; Rizzuti, B.; Dallemagne, P.; Rault, S.; El-Kashef, H.; Saturnino, C.; et al. Benzothienoquinazolinones as new multi-target scaffolds: Dual inhibition of human Topoisomerase I and tubulin polymerization. *Eur. J. Med. Chem.* **2019**, *181*, 111583. [CrossRef] [PubMed]

18. Diaz, P.; Horne, E.; Xu, C.; Hamel, E.; Wagenbach, M.; Petrov, R.R.; Uhlenbruck, B.; Haas, B.; Hothi, P.; Wordeman, L.; et al. Modified carbazoles destabilize microtubules and kill glioblastoma multiform cells. *Eur. J. Med. Chem.* **2018**, *159*, 74–89. [CrossRef]

19. Niu, F.; Liu, Y.; Jing, Z.; Han, G.; Sun, L.; Yan, L.; Zhou, L.; Wu, Y.; Xu, Y.; Hu, L.; et al. Novel carbazole sulfonamide microtubule-destabilizing agents exert potent antitumor activity against esophageal squamous cell carcinoma. *Cancer Lett.* **2018**, *420*, 60–71. [CrossRef]

20. Padmaja, P.; Koteswara Rao, G.; Indrasena, A.; Subba Reddy, B.V.; Patel, N.; Shaik, A.B.; Reddy, N.; Dubey, P.K.; Bhadra, M.P. Synthesis and biological evaluation of novel pyrano[3,2-c]carbazole derivatives as anti-tumor agents inducing apoptosis via tubulin polymerization inhibition. *Org. Biomol. Chem.* **2015**, *13*, 1404–1414. [CrossRef]

21. Bianchi, L.; Giorgi, G.; Maccagno, M.; Petrillo, G.; Scapolla, C.; Tavani, C. An original route to newly-functionalized indoles and carbazoles starting from the ring-opening of nitrothiophenes. *Tetrahedron Lett.* **2012**, *53*, 752–757. [CrossRef]

22. Benzi, A.; Bianchi, L.; Maccagno, M.; Pagano, A.; Petrillo, G.; Tavani, C. Sequential Annulations to Interesting Novel Pyrrolo[3,2-c]carbazoles. *Molecules* **2019**, *24*, 3802. [CrossRef]

23. Benzi, A.; Bianchi, L.; Giorgi, G.; Maccagno, M.; Petrillo, G.; Tavani, C. 2-Aryl-3-Vinyl Substituted Imidazo[1,2-a]pyridines and Fluorescent Electrocyclization Derivatives therefrom. *Chemistryselect* **2020**, *5*, 4552–4558. [CrossRef]

24. Iacopetta, D.; Rosano, C.; Puoci, F.; Parisi, O.I.; Saturnino, C.; Caruso, A.; Longo, P.; Ceramella, J.; Malzert-Freon, A.; Dallemagne, P.; et al. Multifaceted properties of 1,4-dimethylcarbazoles: Focus on trimethoxybenzamide and trimethoxyphenylurea derivatives as novel human topoisomerase II inhibitors. *Eur. J. Pharm. Sci. Off. J. Eur. Fed. Pharm. Sci.* **2017**, *96*, 263–272. [CrossRef] [PubMed]

25. Fazio, A.; Iacopetta, D.; La Torre, C.; Ceramella, J.; Muia, N.; Catalano, A.; Carocci, A.; Sinicropi, M.S. Finding solutions for agricultural wastes: Antioxidant and antitumor properties of pomegranate Akko peel extracts and beta-glucan recovery. *Food Funct.* **2018**, *9*, 6618–6631. [CrossRef] [PubMed]

26. Ceramella, J.; Loizzo, M.R.; Iacopetta, D.; Bonesi, M.; Sicari, V.; Pellicano, T.M.; Saturnino, C.; Malzert-Freon, A.; Tundis, R.; Sinicropi, M.S. Anchusa azurea Mill. (Boraginaceae) aerial parts methanol extract interfering with cytoskeleton organization induces programmed cancer cells death. *Food Funct.* **2019**, *10*, 4280–4290. [CrossRef] [PubMed]

27. Waight, A.B.; Bargsten, K.; Doronina, S.; Steinmetz, M.O.; Sussman, D.; Prota, A.E. Structural Basis of Microtubule Destabilization by Potent Auristatin Anti-Mitotics. *PLoS ONE* **2016**, *11*, e0160890. [CrossRef]

28. Morris, G.M.; Huey, R.; Lindstrom, W.; Sanner, M.F.; Belew, R.K.; Goodsell, D.S.; Olson, A.J. AutoDock4 and AutoDockTools4: Automated docking with selective receptor flexibility. *J. Comput. Chem.* **2009**, *30*, 2785–2791. [CrossRef]

29. Rosano, C.; Lappano, R.; Santolla, M.F.; Ponassi, M.; Donadini, A.; Maggiolini, M. Recent advances in the rationale design of GPER ligands. *Curr. Med. Chem.* **2012**, *19*, 6199–6206. [CrossRef]

30. Iacopetta, D.; Grande, F.; Caruso, A.; Mordocco, R.A.; Plutino, M.R.; Scrivano, L.; Ceramella, J.; Muia, N.; Saturnino, C.; Puoci, F.; et al. New insights for the use of quercetin analogs in cancer treatment. *Future Med. Chem.* **2017**, *9*, 2011–2028. [CrossRef]

31. Cesarini, S.; Spallarossa, A.; Ranise, A.; Schenone, S.; Rosano, C.; La Colla, P.; Sanna, G.; Busonera, B.; Loddo, R. N-acylated and N,N'-diacylated imidazolidine-2-thione derivatives and N,N'-diacylated tetrahydropyrimidine-2(1H)-thione analogues: Synthesis and antiproliferative activity. *Eur. J. Med. Chem.* **2009**, *44*, 1106–1118. [CrossRef]

32. Pettersen, E.F.; Goddard, T.D.; Huang, C.C.; Couch, G.S.; Greenblatt, D.M.; Meng, E.C.; Ferrin, T.E. UCSF Chimera—a visualization system for exploratory research and analysis. *J. Comput. Chem.* **2004**, *25*, 1605–1612. [CrossRef] [PubMed]

33. Bianchi, L.; Maccagno, M.; Petrillo, G.; Sancassan, F.; Spinelli, D.; Tavani, C. 2,3-Dinitro-1,3-butadienes: Versatile building-blocks from the ring-opening of 3,4-dinitrothiophene. In *Targets in Heterocyclic Systems: Chemistry and Properties*, 1st ed.; Attanasi, O.A., Spinelli, D., Eds.; Società Chimica Italiana: Rome, Italy, 2006; Volume 10, pp. 1–23.

34. Bianchi, L.; Maccagno, M.; Petrillo, G.; Rizzato, E.; Sancassan, F.; Severi, E.; Spinelli, D.; Tavani, C.; Viale, M. Versatile nitrobutadienic building-blocks from the ring opening of 2- and 3-nitrothiophenes. In *Targets in Heterocyclic Systems: Chemistry and Properties*, 1st ed.; Attanasi, O.A., Spinelli, D., Eds.; Società Chimica Italiana: Rome, Italy, 2007; Volume 11, pp. 1–20.

35. Petrillo, G.; Benzi, A.; Bianchi, L.; Maccagno, M.; Pagano, A.; Tavani, C.; Spinelli, D. Recent advances in the use of conjugated nitro or dinitro-1,3-butadienes as building-blocks for the synthesis of heterocycles. *Tetrahedron Lett.* **2020**, *61*, 152297. [CrossRef]

36. Fenoglio, C.; Grosso, A.; Petrillo, G.; Boncompagni, E.; Aiello, C.; Cordazzo, C.; Spinelli, D.; Ognio, E.; Mariggio, M.A.; Cassano, A.; et al. A histochemical approach to the evaluation of the in vivo cytotoxicity of the nitrobutadienes (1E,3E)-1,4-bis(1-naphthyl)-2,3-dinitro-1,3-butadiene and methyl (2Z,4E)-2-methylsulfanyl-5-(1-naphthyl)-4-nitro-2,4-pentadienoate in mice liver and kidney. *Anticancer Res.* **2008**, *28*, 813–823. [PubMed]

37. Fontana, A.; Viale, M.; Guernelli, S.; Gasbarri, C.; Rizzato, E.; Maccagno, M.; Petrillo, G.; Aiello, C.; Ferrini, S.; Spinelli, D. Strategies for improving the water solubility of new antitumour nitronaphthylbutadiene derivatives. *Org. Biomol. Chem.* **2010**, *8*, 5674–5681. [CrossRef] [PubMed]

38. Novi, M.; Ottone, M.; Dell'Erba, C.; Barbieri, F.; Chiavarina, B.; Maccagno, M.; Viale, M. 1,4-Bis(1-naphthyl)-2,3-dinitro-1,3-butadiene a novel anticancer compound effective against tumor cell lines characterized by different mechanisms of resistance. *Oncol. Rep.* **2004**, *12*, 91–96. [CrossRef]

39. Petrillo, G.; Fenoglio, C.; Ognio, E.; Aiello, C.; Spinelli, D.; Mariggio, M.A.; Maccagno, M.; Morganti, S.; Cordazzo, C.; Viale, M. Naphthylnitrobutadienes as pharmacologically active molecules: Evaluation of the in vivo antitumour activity. *Investig. New Drugs* **2007**, *25*, 535–544. [CrossRef]

40. Petrillo, G.; Mariggio, M.A.; Aiello, C.; Cordazzo, C.; Fenoglio, C.; Morganti, S.; Croce, M.; Rizzato, E.; Spinelli, D.; Maccagno, M.; et al. Design, synthesis, and in vitro evaluation of new naphthylnitrobutadienes with potential antiproliferative activity: Toward a structure/activity correlation. *Bioorg. Med. Chem.* **2008**, *16*, 240–247. [CrossRef]

41. Petrillo, G.; Tavani, C.; Bianchi, L.; Benzi, A.; Cavalluzzi, M.M.; Salvagno, L.; Quintieri, L.; De Palma, A.; Caputo, L.; Rosato, A.; et al. Densely Functionalized 2-Methylideneazetidines: Evaluation as Antibacterials. *Molecules* **2021**, *26*, 3891. [CrossRef]

42. Tavani, C.; Bianchi, L.; De Palma, A.; Passeri, G.I.; Punzi, G.; Pierri, C.L.; Lovece, A.; Cavalluzzi, M.M.; Franchini, C.; Lentini, G.; et al. Nitro-substituted tetrahydroindolizines and homologs: Design, kinetics, and mechanism of alpha-glucosidase inhibition. *Bioorg. Med. Chem. Lett.* **2017**, *27*, 3980–3986. [CrossRef]

43. Viale, M.; Petrillo, G.; Aiello, C.; Fenoglio, C.; Cordazzo, C.; Mariggio, M.A.; Cassano, A.; Prevosto, C.; Ognio, E.; Maccagno, M.; et al. Synthesis, in vitro activity and in vivo toxicity of the new 2,3-dinitrobutadiene derivative (1E,3E)-1,4-bis(2-naphthyl)-2,3-dinitro-1,3-butadiene. *Pharm. Res.* **2007**, *56*, 318–328. [CrossRef] [PubMed]

44. Viale, M.; Petrillo, G.; Maccagno, M.; Castagnola, P.; Aiello, C.; Cordazzo, C.; Mariggio, M.A.; Jadhav, S.A.; Bianchi, L.; Leto, G.; et al. Sensitivity of different resistant tumour cell lines to the two novel compounds (2Z,4E)-2-methylsulfanyl-5-(1-naphthyl)-4-nitro-2,4-pentadienoate and (1E,3E)-1,4-bis(2-naphthyl)-2,3-dinitro-1,3-butadiene. *Eur. J. Pharm.* **2008**, *588*, 47–51. [CrossRef] [PubMed]

45. Cermak, V.; Dostal, V.; Jelinek, M.; Libusova, L.; Kovar, J.; Rosel, D.; Brabek, J. Microtubule-targeting agents and their impact on cancer treatment. *Eur. J. Cell Biol.* **2020**, *99*, 151075. [CrossRef]

46. Karahalil, B.; Yardim-Akaydin, S.; Nacak Baytas, S. An overview of microtubule targeting agents for cancer therapy. *Arh. Hig. Rada Toksikol.* **2019**, *70*, 160–172. [CrossRef]

47. Tangutur, A.D.; Kumar, D.; Krishna, K.V.; Kantevari, S. Microtubule Targeting Agents as Cancer Chemotherapeutics: An Overview of Molecular Hybrids as Stabilizing and Destabilizing Agents. *Curr. Top. Med. Chem.* **2017**, *17*, 2523–2537. [CrossRef]

48. Krause, W. Resistance to anti-tubulin agents: From vinca alkaloids to epothilones. *Cancer Drug Resist.* **2019**, *2*, 82–106. [CrossRef]

49. Zhou, J.; Giannakakou, P. Targeting microtubules for cancer chemotherapy. *Curr. Med. Chem. Anticancer Agents* **2005**, *5*, 65–71. [CrossRef]

50. Cong, H.; Zhao, X.; Castle, B.T.; Pomeroy, E.J.; Zhou, B.; Lee, J.; Wang, Y.; Bian, T.; Miao, Z.; Zhang, W.; et al. An Indole-Chalcone Inhibits Multidrug-Resistant Cancer Cell Growth by Targeting Microtubules. *Mol. Pharm.* **2018**, *15*, 3892–3900. [CrossRef]
51. Mukhtar, E.; Adhami, V.M.; Mukhtar, H. Targeting microtubules by natural agents for cancer therapy. *Mol. Cancer* **2014**, *13*, 275–284. [CrossRef]
52. Raja, V.J.; Lim, K.H.; Leong, C.O.; Kam, T.S.; Bradshaw, T.D. Novel antitumour indole alkaloid, Jerantinine A, evokes potent G2/M cell cycle arrest targeting microtubules. *Investig. New Drugs* **2014**, *32*, 838–850. [CrossRef] [PubMed]
53. Tantak, M.P.; Malik, M.; Klingler, L.; Olson, Z.; Kumar, A.; Sadana, R.; Kumar, D. Indolyl-alpha-keto-1,3,4-oxadiazoles: Synthesis, anti-cell proliferation activity, and inhibition of tubulin polymerization. *Bioorg. Med. Chem. Lett.* **2021**, *37*, 127842. [CrossRef] [PubMed]
54. Wang, G.; Liu, W.; Gong, Z.; Huang, Y.; Li, Y.; Peng, Z. Design, synthesis, biological evaluation and molecular docking studies of new chalcone derivatives containing diaryl ether moiety as potential anticancer agents and tubulin polymerization inhibitors. *Bioorg. Chem.* **2020**, *95*, 103565. [CrossRef] [PubMed]

*Review*

# Carbazoles: Role and Functions in Fighting Diabetes

**Fedora Grande** [1,†], **Giuseppina Ioele** [1,†], **Anna Caruso** [1,2,*], **Maria Antonietta Occhiuzzi** [1], **Hussein El-Kashef** [3], **Carmela Saturnino** [2] **and Maria Stefania Sinicropi** [1]

1    Department of Pharmacy, Health and Nutritional Sciences, University of Calabria, 87036 Arcavacata di Rende, Italy
2    Department of Science, University of Basilicata, 85100 Potenza, Italy
3    Faculty of Science, Assiut University, Assiut 71516, Egypt
*    Correspondence: anna.caruso@unical.it or anna.caruso@unibas.it
†    These authors contributed equally to this work.

**Abstract:** Carbazole derivatives have gained a lot of attention in medicinal chemistry over the last few decades due to their wide range of biological and pharmacological properties, including antibacterial, antitumor, antioxidant, and anti-inflammatory activities. The therapeutic potential of natural, semi-synthetic or synthetic carbazole-containing molecules has expanded considerably owing to their role in the pathogenesis and development of diabetes. Several studies have demonstrated the ability of carbazole derivatives to reduce oxidative stress, block adrenergic hyperactivation, prevent damage to pancreatic cells and modulate carbohydrate metabolism. In this survey, we summarize the latest advances in the synthetic and natural carbazole-containing compounds involved in diabetes pathways.

**Keywords:** carbazoles; alkaloids; diabetes mellitus; antihyperglycemic agents

**Citation:** Grande, F.; Ioele, G.; Caruso, A.; Occhiuzzi, M.A.; El-Kashef, H.; Saturnino, C.; Sinicropi, M.S. Carbazoles: Role and Functions in Fighting Diabetes. *Appl. Sci.* **2023**, *13*, 349. https://doi.org/10.3390/app13010349

Academic Editor: Monica Gallo

Received: 7 November 2022
Revised: 20 December 2022
Accepted: 22 December 2022
Published: 27 December 2022



## 1. Introduction

The term "diabetes mellitus" (DM) refers to a group of chronic metabolic disorders characterized by hyperglycemia due to insufficient insulin secretion (type 1 DM) or to an inadequate response of the body to the action of this hormone (type 2 DM) [1]. Since a constant increase in glucose concentration in the blood can lead to serious consequences for several organs, including the eyes, kidneys, cardiovascular system, and central nervous system (CNS) [2,3] monitoring glycemia is of primary importance [4,5]

An effective approach to reduce postprandial blood glucose concentration is to inhibit the enzyme α-glucosidase, which catalyzes the final stage of the digestion process of carbohydrates to monosaccharides. These latter sugars are then absorbed and enter the bloodstream, thus increasing blood glucose levels [6]. The damages caused by hyperglycemia are related to the alteration of several physiological pathways. Several studies have shown, for example, that high levels of glucose, fatty acids, and insulin in the blood promote the production of reactive oxygen species (ROS), which cause direct DNA damage and promote insulin resistance [7].

Furthermore, by overstimulating the IGF-1 factor or by activating different kinase proteins, hyperglycemia favors cellular hyperproliferation and consequently the onset of cancer [8–10]. As a result of scientific advances, a wide range of anti-diabetic drugs are now available, improving the quality and length of life in type 2 diabetic patients.

These drugs, which belong to different chemical classes, can lower blood glucose levels through a variety of mechanisms (see Table 1 and Figure 1).

**Table 1.** Classes of antidiabetic medications.

| DRUG CLASS | MECHANISM OF ACTION | MORE COMMON SIDE EFFECTS AND DISADVANTAGES |
|---|---|---|
| **BIGUANIDES:**<br>**Metformin** | AMP-Kinase activation;<br>Decrease in hepatic glucose production. | Gastrointestinal side effects (diarrhea, abdominal cramping);<br>Lactic acidosis risk;<br>Vit B12 deficiency;<br>Hypoxia;<br>Dehydration. |
| **SULPHONYLUREAS:**<br>**Glibenclamide**<br>**Gliclazide**<br>**Glipizide**<br>**Glimepiride** | Closure of KATP channels on $\beta$ cell plasma membranes;<br>Increase in insulin secretion. | Hypoglycemia;<br>Weight gain. |
| **GLINIDES (MEGLITINIDES):**<br>**Repaglinide**<br>**Nateglinide** | Closure of KATP channels on $\beta$ cell plasma membranes;<br>Increase in insulin secretion. | Hypoglycemia;<br>Weight gain. |
| **$\alpha$-GLUCOSIDASE INHIBITORS:**<br>**Acarbose**<br>**Miglitol** | Inhibition of intestinal $\alpha$-glucosidase;<br>Intestinal carbohydrate digestion and absorption reduction. | Generally modest HbA1c efficacy;<br>Gastrointestinal side effects (flatulence, diarrhea). |
| **THIAZOLIDINEDIONES (TZDs):**<br>**Pioglitazone**<br>**Rosiglitazone**<br>**Lobeglitazone** | Nuclear transcription factor PPAR-$\gamma$ activation;<br>Increase in insulin sensitivity. | Weight gain;<br>Edema;<br>Heart failure;<br>Bone fractures. |
| **DPP4 INHIBITORS (GLIPTINS):**<br>**Sitagliptin**<br>**Vildagliptin**<br>**Saxagliptin**<br>**Alogliptin**<br>**Alogliptin**<br>**Linagliptin** | DPP-4 inhibition;<br>Increase in postprandial active incretin (GLP-1, GIP) concentrations;<br>Increase in glucose-dependent insulin secretion;<br>Decrease in glucose-dependent glucagon secretion. | Generally modest HbA1c efficacy;<br>Urticaria;<br>Angioedema. |
| **GLP-1 AGONISTS:**<br>**Exenatide**<br>**Liraglutide**<br>**Lixisenatide**<br>**Dulaglutide** | GLP-1 receptors activation;<br>Increase in glucose-dependent insulin secretion;<br>Decrease in glucose-dependent glucagon secretion.;<br>Slowing of gastric emptying;<br>Satiety increase. | Gastrointestinal side effects (nausea/vomiting);<br>Pancreatitis;<br>C-cell hyperplasia. |
| **SGLT2 INHIBITORS:**<br>**Dapagliflozin**<br>**Empagliflozin**<br>**Canagliflozin** | Block of sodium/glucose cotransporter 2 (SGLT2) in renal tubules<br>Reduction of glucose reabsorption in the kidney;<br>Decrease in serum blood glucose level. | Renal failure;<br>Increased risk of genital and urinary trac fungal infection;<br>Increased risk of euglycemic diabetic ketoacidosis. |
| **AMYLIN ANALOGUES**<br>**Pramlintide** | Amylin receptors activation;<br>Glucagon secretion reduction;<br>Slowing of gastric emptying;<br>Satiety increase. | Generally modest HbA1c efficacy;<br>Gastrointestinal side effects<br>Hypoglycemia. |

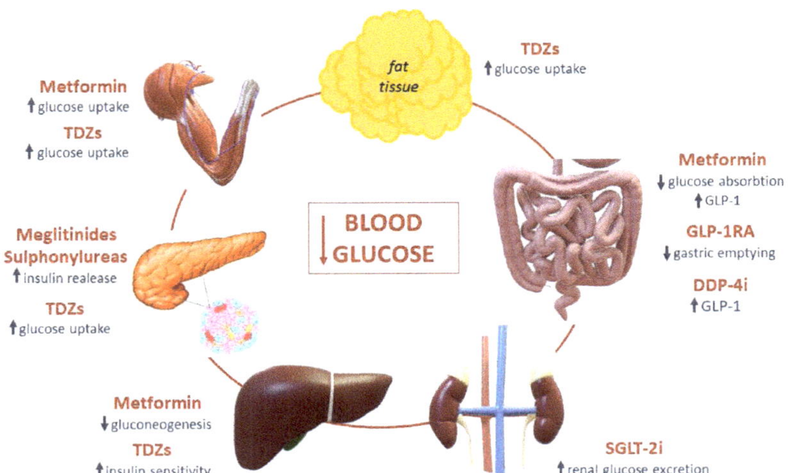



**Figure 1.** Mechanisms of action of antidiabetic drugs.

Metformin, one of the most commonly used oral hypoglycemic drugs, acts, for example, by increasing the utilization of glucose in peripheral tissues and by decreasing hepatic glucose production. Sulfonylureas, on the other hand, are able to modulate blood glucose levels by stimulating insulin secretion from pancreatic beta cells. Thiazolidinediones (TZDs) binds the peroxisome proliferator-activated receptor (PPAR) gamma, a transcription factor that regulates the expression of specific genes, and improves the sensitivity of fatty tissue, skeletal muscles and liver to insulin. More recently, new classes of drugs able to control hyperglycemia have been introduced. Among these, incretins, DPP-4 resistant GLP-1 analogs, and DPP-4 inhibitors play an important role; they are able to promote an increase in endogenous GLP-1 levels and stimulate pancreatic cells to release insulin. SGLT2 inhibitors increase glucose excretion in the urine by preventing glucose reabsorption in the renal tubules. Despite the therapeutic goals achieved by this diverse range of diabetes drugs, their clinical use is associated with undesirable side effects (summarized in Table 1), including dangerous hypoglycemic episodes [11]. Therefore, the identification of new molecules that preserve hypoglycemic activity while having reduced toxicity remains a goal of the scientific community. In this context, carbazole derivatives could represent valid therapeutic alternatives. Carbazole (I) (Figure 2) is a tricyclic heterocycle alkaloid consisting of two benzene rings fused on both sides of a pyrrole ring [12–15].

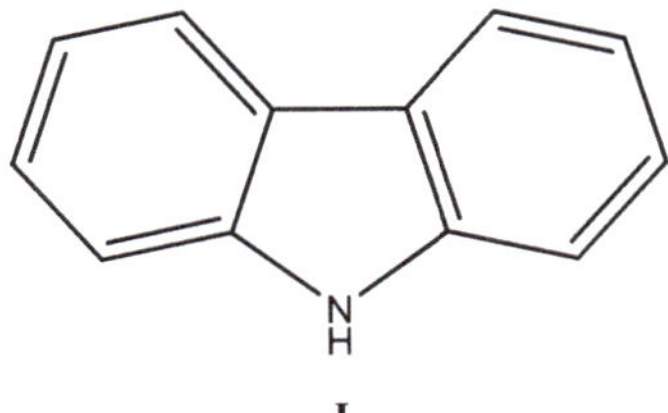

**I**

**Figure 2.** Structure of carbazole (I).

The first naturally occurring carbazole was isolated from Murraya *koenigii Spreng*. Murraya *euchrestifolia* has been found to be a rich source of carbazole alkaloids, providing a variety of novel structures. Some bioactive carbazole alkaloids have also been obtained from other sources such as actinomycetes, blue-green algae and mammalian systems [16,17].

A large number of studies reported in the literature were focused on the biological properties of carbazole and its derivatives, such as antimicrobial, anti-inflammatory, anti-tumor, antioxidant, antiepileptic, antihistamine, antidiarrheal, analgesic, neuroprotective, and inhibiting properties of pancreatic lipase [18–21].

In particular, some compounds belonging to this chemical class have shown promising activities in controlling glucose metabolism in hypertensive patients with type 2 DM and in improving insulin sensitivity [22]. In this review, we summarize the latest advances in the potential role of carbazole derivatives either in the prevention of or in the treatment of diabetes. Overall, the data herein reported could provide an important resource for the development of novel, efficient, and safe agents for innovative diabetes treatment.

## 2. Carbazole Derivatives in the Pathogenesis of Diabetes

Over the years, the academic community has delved into the chemical and pharmacological properties of small heterocyclic molecules, including those with a carbazole-based structure [23–26]. Growing interest in these derivatives has resulted in the development of various synthetic routes for their preparation as well as their extraction from various plants. Several carbazole-containing molecules have found application in the pharmaceutical field. In particular, the studies included in this review showed that certain derivatives are promising antidiabetic agents or may otherwise play an important role in the pathogenesis of diabetes. Data reported in the literature suggest that carbazoles are able to modulate glucose metabolism, block adrenergic hyperactivation, prevent damage to β cells of the pancreas, inhibit inflammatory and oxidative mediators, control the cryptochrome or inhibit $\alpha$-glucosidase (Table 2). Therefore, carbazole represents a versatile scaffold for the preparation of biologically active derivatives, useful in the treatment of diabetes [6,7,27–46].

**Table 2.** Carbazole derivatives in the pathogenesis of diabetes.

| Compound | Name | Biological Activity | References |
|---|---|---|---|
| **1** (Synthetic compound) | Carvedilol (1-(9*H*-carbazol-4-yloxy)-3-[2-(2-methoxyphenoxy)ethylamino]propan-2-ol) | Reduction of insulin resistance by sensitizing insulin receptors and inhibiting the sympathetic nervous system Beneficial effects on left ventricular function, resting and exercise hemodynamics and clinical conditions Stimulation of endothelial NO production Beneficial effects on endothelial dysfunction caused by oxidative stress Long-term benefits on glucose metabolism Blockade of adrenergic hyperactivation Prevention of pancreatic $\beta$-cell damage Inhibition of inflammatory and oxidative mediators | [7,27–33] |
| **2** | Hydroxyphenyl-carvedilol (OHC) 4-[2-[[3-(9*H*-Carbaz-ol-4-yloxy)-2-hydroxypro-pyl]amino]ethoxy]-3-methoxyphenol | Metabolite of carvedilol | [30] |
| **3** | *O*-Desmethylcarvedilol (DMC) 2-[2-[[3-(9*H*-Carbaz-ol-4-yloxy)-2-hydroxypro-pyl]amino]ethoxy]phenol | Metabolite of carvedilol | [30] |
| **4** (Synthetic compound) | 1-((5,6-Di(furan-2-yl)-1,2,4-triazin-3-yl)thio)-3-(3,6-dibromo-9*H*-carbazol-9-yl)propan-2-ol | $\alpha$-Glucosidase inhibition (IC$_{50}$ = 4.27 $\pm$ 0.07 $\mu$M) | [6,34] |
| **5** (Synthetic compound) | 2-(3-(9*H*-Carbazol-9-yl)-2-hydroxypropyl)isothiazoline-1,1-dioxide | Cryptochrome modulator | [35] |

**Table 2.** *Cont.*

| Compound | Name | Biological Activity | References |
|---|---|---|---|
| **6** (Synthetic compound) | 1-(3-(3,6-Difluoro-9*H*-carbazol-9-yl)-2-hydroxypropyl)imidazolidin-2-one | Cryptochrome modulator | [22] |
| **7** (Synthetic compound) | 2-(4-((9*H*-Carbazol-9-yl)methyl)-1*H*-1,2,3-triazol-1-yl)-1-(3-bromo-4-hydroxyphenyl) ethanone | α-Glucosidase inhibition ($IC_{50}$ = 1.0 ± 0.057 μM) | [36] |
| **8** (Synthetic compound) | 9-((1-(Pyridin-3-yl-methyl)-1*H*-1,2,3-triazol-4-yl) methyl)-9*H*-carbazole | α-Glucosidase inhibition ($IC_{50}$ = 0.8 ± 0.01 μM) | [36] |
| **9** (Synthetic compound) | 6-(Benzyloxy)-9-(4-chlorobenzoyl)-2,3,4,9-tetrahydro-1*H*-carbazole-3-carboxylic acid | Hypoglycemic effect via the AMPK pathway | [37] |
| **10** (Synthetic compound) | Ethyl 8-(benzyloxy)-5-(4-chlorobenzoyl)-7-fluoro-3,4-dihydro-1*H*-pyrido[4,3-*b*]indole-2(5*H*)-carboxylate | Hypoglycemic effect via the AMPK pathway | [38] |
| **11** (Naturalcompound) | Mahanine 3,5-Dimethyl-3-(4-methylpent-3-enyl)-11*H*-pyrano[3,2-*a*]carbazol-9-ol | Prevention of insulin resistance due to lipid induced signaling defects α-Glucosidase inhibition ($IC_{50}$ = 21.4 ± 0.4 μM) Antioxidant properties Increase in the translocation of GLUT4 protein from intracellular vesicles into the plasma fraction and glucose uptake through the activation of phosphorylation of Akt | [39–41] |
| **12** (Naturalcompound) | Bisgerayafoline D 3,3′-Bis((*E*)-3,7-dimethylocta-2,6-dien-1-yl)-9′-methoxy-3,3′,5,5′-tetramethyl-3,3′,11,11′-tetrahydro-[9,10′-bipyrano[3,2-*a*]carbazol]-10-ol | Antioxidant and α-glucosidase properties | [40] |
| **13** (Naturalcompound) | Bismahanimbinol 3,3′,5,5′-Tetramethyl-3,3′-bis(4-methylpent-3-en-1-yl)-3,3′,11,11′-tetrahydro-[9,10′-bipyrano[3,2-*a*]carbazol]-8-ol | Antioxidant and α-glucosidase properties | [40] |
| **14** (Naturalcompound) | Bispyrayafoline 3,3′,5,5′-Tetramethyl-3,3′-bis(4-methylpent-3-en-1-yl)-3,3′,11,11′-tetrahydro-[10,10′-bipyrano[3,2-*a*]carbazole]-9,9′-diol | Antioxidant and α-glucosidase properties | [40] |
| **15** (Naturalcompound) | *O*-Methyl mahanine 9-Methoxy-3,5-dimethyl-3-(4-methylpent-3-en-1-yl)-3,11-dihydropyrano[3,2-*a*]carbazole | Antioxidant and α-glucosidase properties | [40] |
| **16** (Naturalcompound) | *O*-Methyl mukonal Koenimbine 8-Methoxy-3,3,5-trimethyl-11*H*-pyrano[3,2-*a*]carbazole | Antioxidant and α-glucosidase properties Antidiabetic activity in L6-GLUT4 myc myotubes | [40,42] |
| **17** (Naturalcompound) | *O*-Methylmurrayamine A 9-Methoxy-3,3,5-trimethyl-11*H*-pyrano[3,2-*a*]carbazole | Antidiabetic activity in L6-GLUT4 myc myotubes Decrease in blood glucose profile | [42] |

**Table 2.** *Cont.*

| Compound | Name | Biological Activity | References |
|---|---|---|---|
| **18** (Naturalcompound) | Koenidine 8,9-Dimethoxy-3,3,5-trimethyl-11*H*-pyrano [3,2-*a*]carbazole | Antidiabetic activity in L6-GLUT4myc myotubes Increase in insulin sensitivity and progressive reduction of blood glucose level | [42] |
| **19** (Naturalcompound) | Mahanimbine 3,5-Dimethyl-3-(4-methylpent-3-enyl)-11*H*-pyrano[3,2-*a*]carbazole | Antidiabetic activity in L6-GLUT4myc myotubes | [42] |
| **20** (Naturalcompound) | Murrayazoline (14*R*,17*S*,19*S*)-3,13,13,17-Tetramethyl-21-oxa-12-azahexacyclo[10.7.1.1$^{2,17}$.0$^{5,20}$.0$^{6,11}$.0$^{14,19}$]henicosa-1,3,5(20),6,8,10-hexaene | Antidiabetic activity in L6-GLUT4myc myotubes | [42] |
| **21** (Synthetic compound) | (*S*)-3-(4-(2-(9*H*-Carbazol-9-yl)ethoxy)phenyl)-2-ethoxypropanoic acid | Improve of the insulin sensitivity Activity on PPARR and PPARγ | [43] |
| **22** (Synthetic compound) | Chiglitazar (2*S*)-3-[4-(2-Carbazol-9-ylethoxy)phenyl]-2-[2-(4-fluorobenzoyl)anilino]propanoic acid | Reduction of glycosylated hemoglobin A1c (HbA1c) Hypoglycemic effect Increase in insulin sensitivity Reduction of triglycerides | [44–46] |

## 2.1. Carvedilol

The carbazole derivative **1** (carvedilol, 1-(9*H*-carbazol-4-yloxy)-3-[2-(2-methoxyphenoxy) ethylamino]propan-2-ol) (Figure 3), synthesized by the reaction of 4-(oxirane 2-ylmethoxy)-9*H*-carbazole with 2-(2-methoxyphenoxy)ethylamine [47], was discovered in a study devoted to discovering agents useful in the treatment of congestive heart failure. The interaction of carvedilol (**1**) with beta-adrenergic receptors is stereospecific, in a similar way that was demonstrated for carazolol, a veterinary medicine drug used to reduce stress in animals during transport [48]. This optically active compound exhibits activity as a selective beta-adrenergic receptor blocker such as the *S* (−) enantiomer, whereas the *R* (+) enantiomer is capable of antagonizing both alpha-1 and beta-adrenergic receptors. These characteristics make this molecule more effective than other traditional beta-blockers in the treatment of heart failure and in the regulation of myocardial functions. Furthermore, unlike other beta-blockers, it does not cause alterations in the metabolism of carbohydrates and lipids [49].

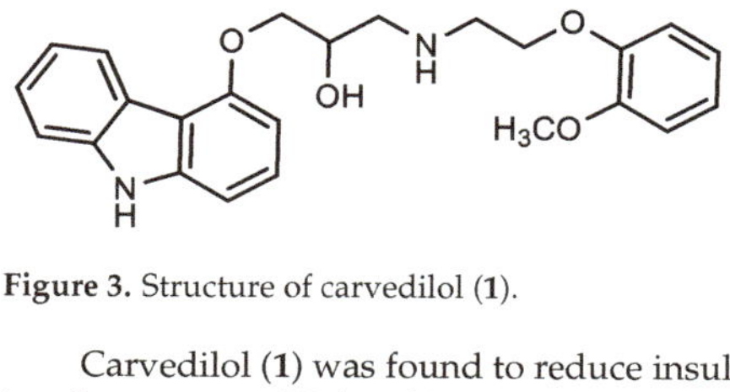

**Figure 3.** Structure of carvedilol (**1**).

Carvedilol (**1**) was found to reduce insulin resistance by increasing the sensitivity of insulin receptors (IRs) and by reducing the activity of the sympathetic nervous system [50]. The adrenergic system, in part through beta-2 adrenergic receptors, regulates glucose and lipid metabolism in the liver, adipose tissue, and skeletal muscle.

Taking into consideration all these properties, several studies were focused on the evaluation of the efficacy of carvedilol (**1**) in preventing heart complications in patients with diabetes [7,51–53]. In an earlier study on the efficacy and tolerability of long-term administration of carvedilol (**1**), Nodari et al. [27] evaluated the drug's effects in patients with chronic heart failure, a number of whom were also diabetic. Treatment with increasing doses of carvedilol was well tolerated by all treated patients, showing positive effects on left ventricular function and on resting and exercise hemodynamics. Furthermore, unlike traditional beta-blockers, it did not affect insulin sensitivity and glycemic values in diabetic patients treated with oral hypoglycemic agents or insulin [27,54,55].

In a separate study, carvedilol (**1**) was found to be more effective than metoprolol, one of the most prescribed selective beta-1 adrenergic antagonists, in controlling glucose metabolism in diabetic hypertensive patients without affecting insulin resistance [51,54]. This was most likely due to carvedilol's ability to also block beta-2 and alpha-1 receptors [56]. Unlike patients treated with metoprolol, who gained weight after 2 months of treatment, no significant change in body weight was observed in those treated with carvedilol (**1**). Long-term carvedilol treatment may have beneficial effects on endothelial dysfunction caused by oxidative stress in patients with type 2 DM. This could be due to the antioxidant properties of the compound, which acts as a radical scavenger, and to its capability to modulate endothelial NO production [28,57].

Vardeny et al., in 2012 [29], found that in hypertensive patients treated with metoprolol succinate, after six months, insulin levels increased by approximately 36%, whereas a decrease of about 10% was observed in those treated with carvedilol (**1**) ($p = 0.015$).

Many studies have discovered a relation between $\beta$2-AR polymorphisms and metabolic disorders, including hypertriglyceridemia, insulin resistance, and obesity [58–60]. In particular, two common $\beta$2-AR single-nucleotide polymorphisms, Arg16Gly and Gln27Glu, have been shown to be responsible for altered receptor function and, consequently, for an inadequate response to $\beta$-blocker treatment.

Farhat et al., in 2019, demonstrated that low-dose carvedilol (**1**) can efficiently prevent hypoglycemic crises in rats treated with $\beta$-blockers. This study showed that rats with recurrent episodes of hypoglycemia treated with carvedilol required less exogenous glucose during the hypoglycemic clamp, thus supporting use of the drug to preserve hormonal responses to hypoglycemia. Furthermore, carvedilol treatment had no effect on plasma lactate levels, indicating that the drug can prevent some of the central adaptations that occur during recurrent hypoglycemic episodes [61].

In 2017, Nardotto et al. [30] conducted a study aimed at evaluating the pharmacokinetic profile of the enantiomers of carvedilol and the corresponding metabolites, hydroxyphenylcarvedilol (OHC) (**2**) and O-desmethylcarvedilol (DMC) (**3**) (Figure 4), in healthy volunteers and in patients with type 2 DM but with good glycemic control.

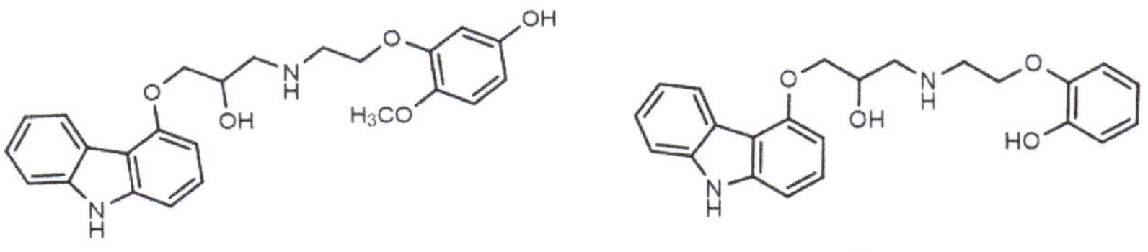

**Figure 4.** Structures of: hydroxyphenylcarvedilol (OHC) (**2**) and O-desmethylcarvedilol (DMC) (**3**).

For this purpose, thirteen healthy subjects were recruited and divided into two groups: the first group underwent treatment with carvedilol as a single agent, while the second one received a treatment of carvedilol (**1**) in combination with glibenclamide and metformin. In addition, patients with type 2 DM were enrolled and were treated with carvedilol as a single agent. The results of this study demonstrated that the pharmacokinetic profile of the carvedilol enantiomers did not differ between each other, even in the patients treated with glibenclamide and metformin.

In a separate study, the same authors, using an integrated population pharmacokinetic modeling approach, confirmed that carvedilol does not induce insulin resistance or worsen glycemic control in diabetic hypertensive patients [31]. More recently, Nguyen et al. conducted a study to evaluate the effects of carvedilol (**1**) treatment in subjects with obesity induced by a high-fat diet (HFD). The findings not only confirmed the improvement in glucose tolerance and insulin sensitivity, but also indicated that these effects are related to the suppression of hepatic glucose overproduction and to the enhancement of the muscle insulin signaling pathway. Hence, as a result of the drug's ability to block adrenergic hyperactivation, long-term treatment with carvedilol (**1**) may provide significant benefits in obese subjects [32,60,61].

The efficacy of carvedilol (**1**) is, however, limited due to its poor bioavailability. Therefore, various studies have been conducted in order to formulate the drug in transport systems capable of guaranteeing the desired therapeutic effect without increasing the dose to be administered. Since a high insulin concentration significantly increases DNA fragmentation, Farahani-Zangaraki et al., in 2021 [7], used the Comet test to assess the genoprotective effects of carvediol included into niosomes against supraphysiological insulin levels in human umbilical vein endothelial cells (HUVEC) [62]. As a result, incorporating carvedilol (**1**) into the nanoparticles increased efficiency by about fivefold. In particular, treatment of HUVEC cells with niosome-carvedilol (**1**) nanoparticles 24 h before insulin administration resulted in a significant decrease in DNA fragmentation related to the insulin-treated group, confirming the better genoprotective effect of the drug loaded in the niosomes compared to the free drug. Treatment of HUVEC cells with niosome carvedilol (**1**) nanoparticles 24 h before insulin administration decreases DNA fragmentation compared to the insulin-treated group, confirming the greater genoprotective effect of the drug loaded in the niosomes compared to the free drug [7].

Recently, carvedilol (**1**) had also been investigated as an alternative therapeutic strategy for the treatment of type 1 DM. In an in vivo experimental model of type 1 DM, Amirshahrokhi and Zohouri [33] demonstrated carvedilol's protective effect against pancreatic $\beta$-cell damage, confirming the drug's ability to significantly reduce blood glucose levels, weight loss, and insulitis in pancreatic tissue, as well as the onset of diabetes. Significant increases in the antioxidants glutathione (GSH), superoxide dismutase (SOD), and catalase were observed in pancreatic tissue from carvedilol-treated mice. On the other hand, a reduction in malondialdehyde (MDA), nitric oxide (NO), and myeloperoxidase (MPO) levels was also observed. In addition, the treatment promoted an appreciable reduction in the $\beta$-cell damage in pancreatic tissue as well as in the expression of inflammatory modulators including the nuclear factor kappa B (NF-$\kappa$B), cyclooxygenase-2 (COX-2) and inducible nitric oxide synthase (iN-OS). Similarly, a reduction of proinflammatory cytokines tumor necrosis factor (TNF)-$\alpha$, interleukin (IL)-1$\beta$, IL-6 IL-12, IL-17, interferon (IFN)-$\gamma$, and chemokine had been also observed, while the expression of anti-inflammatory cytokine IL-10 increased [33].

Despite that the experimental evidence points to carvedilol as having a decidedly positive effect in the treatment of diabetes, more research into the drug's mechanism of action is required.

### 2.2. 1,2,4-Triazine-Carbazoles

As already mentioned, $\alpha$-glucosidase represents a valuable target for the treatment of type 2 DM. As a result, numerous efforts are being made by the scientific community to identify new compounds capable of inhibiting this enzyme in order to delay carbohydrate digestion and glucose absorption, thus leading to a reduction in postprandial glycemia values [63]. Accordingly, in 2016, Wang et al. [6] developed a small series of triazines, used as different starting substitute carbazoles, in order to identify novel $\alpha$-glucosidase inhibitors.

A number of the newly synthesized compounds showed noteworthy activity against the enzyme. In particular, compound **4** (1-((5,6-di(furan-2-yl) -1,2,4-triazin-3-yl)thio)-3-(3,6-dibromo-9H-carbazol-9-yl)propan-2-ol)) (Figure 5) was found to be the most promising, with $IC_{50}$ values of $4.27 \pm 0.07$ µM, significantly higher than the control drug acarbose ($IC_{50} = 995.55 \pm 2.71$ µM).

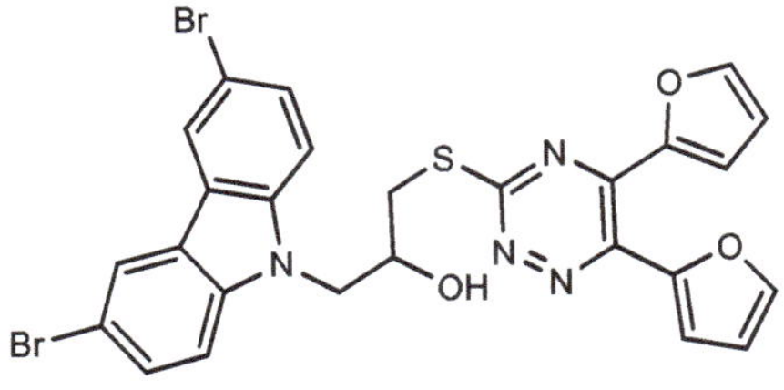

**Figure 5.** Structure of 1-((5,6-di(furan-2-yl)-1,2,4-triazin-3-yl)thio)-3-(3,6-dibromo-9H-carbazol-9-yl)propan-2-ol (**4**).

Molecular docking experiments confirmed its capability to interact with the active site of the enzyme, and furthermore, the kinetic analysis revealed that **4** acts as a non-competitive inhibitor. In particular, the carbazole ring of **4** forms arene-cation interactions with residues Arg-439 and Arg-312, respectively, and the furan ring interacts with residues Leu-176, Phe-157, Phe-177 and Pro-240. Overall, the findings of this study could represent a starting point for the future development of efficient α-glucosidase inhibitors.

### 2.3. Sulfonamide Carbazole and Carbazole-Containing Cyclic Urea

Variants of the cryptochrome (Cry) gene have been demonstrated to be correlated to the onset of type 2 DM and insulin resistance, as it is involved in glucose homeostasis, control of β-cell function liver lipid content, and the circadian system of mammals [22,64,65].

In order to identify compounds capable of modulating the cryptochrome activity, Humphries et al. in 2016 [35] carried out a systematic SAR study that led to a series of sulfonamide and sulfamide carbazole-based derivatives. Among these compounds, cyclic sulfonamide, 2-(3-(9*H*-carbazol-9-yl)-2-hydroxypropyl)isothiazoline-1,1-dioxide (**5**) (Figure 5) was identified as the first small molecule, orally bioavailable, active as a cryptochrome modulator in an in vivo model of type 2 diabetes. The efficacy of this compound was evaluated in mice with diet-induced obesity (DIO), using rosiglitazone as a positive control.

The results obtained by an oral glucose tolerance test confirmed a significantly improved glucose clearance [35]. In 2018, the same research team discovered a new class of carbazole-containing amides and ureas as cryptochrome modulators [22]. In particular, compound 1-(3-(3,6-difluoro-9*H*-carbazol-9-yl)-2-hydroxypropyl)imidazolidin-2-one (**6**) (Figure 6) resulted as the most promising derivative, implying its potential use in the treatment of metabolic disorders including type 2 DM [22,66–69].

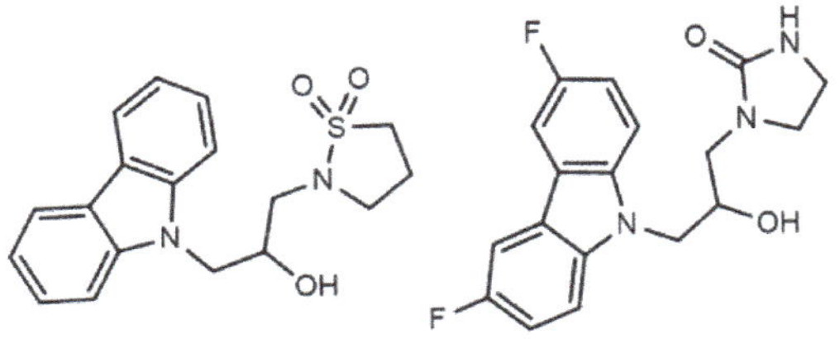

**Figure 6.** Structures of: 2-(3-(9*H*-carbazol-9-yl)-2-hydroxypropyl)isothiazoline-1,1-dioxide (**5**) and 1-(3-(3,6-difluoro-9*H*-carbazol-9-yl)-2-hydroxypropyl)imidazolidin-2-one (**6**).

### 2.4. Carbazole Triazoles

Recently, a series of derivatives of carbazole linked to a variously substituted triazole have been developed and tested as α-glucosidase inhibitors [36]. The synthesis of these compounds was carried out by a click reaction using *N*-propargyl-9*H*-carbazole, acetophenone azide, and a suitable heterocycle as starting reagents. Almost all of the synthesized compounds inhibited the target enzyme more efficiently than acarbose, used as a positive control.

Compounds 2-(4-((9*H*-carbazol-9-yl)methyl)-1*H*-1,2,3-triazol-1-yl)-1-(3-bromo-4-hydroxyphenyl) ethanone (**7**) and 9-((1-(pyridin-3-yl-methyl)-1*H*-1,2,3-triazol-4-yl) methyl)-9*H*-carbazole (**8**) (Figure 7), with IC$_{50}$ values of 1.0 and 0.8 μM, respectively, resulted as the most promising and thus represent suitable lead compounds for the development of innovative non-sugar derivative antidiabetic agents.

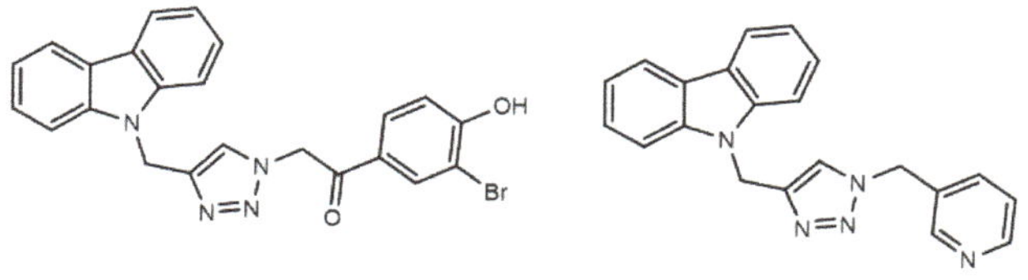

**Figure 7.** Structures of: 2-(4-((9*H*-carbazol-9-yl)methyl)-1*H*-1,2,3-triazol-1-yl)-1-(3-bromo-4-hydroxyphenyl) ethanone (**7**) and 9-((1-(pyridin-3-yl-methyl)-1*H*-1,2,3-triazol-4-yl) methyl)-9*H*-carbazole (**8**).

### 2.5. Tetrahydrocarbazole Derivatives

Zhang et al., in 2018 [37], developed a new series of tetrahydrocarbazoles and tested them in an in vitro assay on human hepatoma cell lines (HepG2) to assess their hypoglycemic activity. Several of the compounds tested exhibited significant activity, with carboxyl (**9**), 6-(benzyloxy)-9-(chlorobenzoyl)-2,3,4,9-tetrahydro-1*H*-carbazole-3-carboxylic acid (Figure 8) being the most promising. This compound was thus selected for further in vivo experiments. The results obtained confirmed that compound **9** has comparable hypoglycemic properties to that of pioglitazone and causes less weight gain compared to the drug in clinical use.

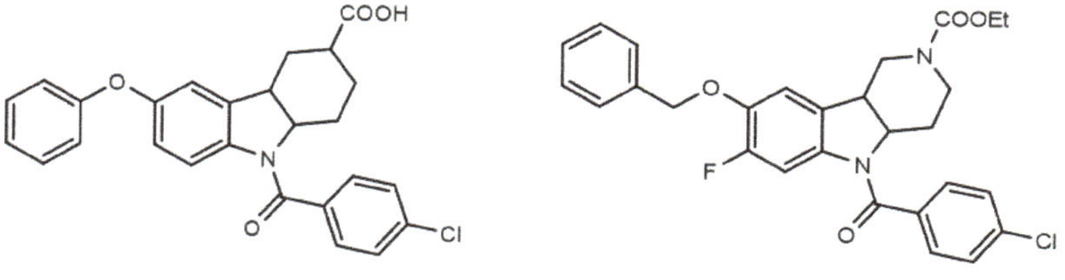

**Figure 8.** Structures of: 6-(benzyloxy)-9-(chlorobenzoyl)-2,3,4,9-tetrahydro-1*H*-carbazole-3-carboxylic acid (**9**) and ethyl 8-(benzyloxy)-5-(4-chlorobenzoyl)-7-fluoro-3,4-dihydro-1*H*-pyrido[4,3-*b*]indole-2(5*H*)-carboxylate (**10**).

A recent study [42], reported by the same above research group, led to the identification of an azatetrahydrocarbazole derivative, namely ethyl 8-(benzyloxy)-5-(4-chlorobenzoyl)-7-fluoro-3,4-dihydro-1*H*-pyrido[4,3-b]indole-2(5*H*)-carboxylate (**10**) (Figure 8). This derivative showed hypoglycemic activity, in cell assays, approximately 1.2 times higher than that of metformin, used as a positive control. Studies aimed at elucidating the mechanism of action responsible for the hypoglycemic activity of these compounds have suggested involvement of the AMP-activated protein kinase (AMPK). The AMPK-mediated pathway plays a key role in controlling the energy and metabolic homeostasis of cells, acting as a sensor for the cellular energy status [70,71].

Compound **10** also showed appreciable chemical stability in gastrointestinal fluids and plasma in in vivo studies, which also confirmed a good tolerance of the compound after oral administration [38]. The promising hypoglycemic, pharmacokinetic, and pharmacodynamic properties of compound **10** validate the potential of carbazole as a versatile scaffold for the preparation of derivatives with improved bioactivities that can be useful in a variety of therapeutic applications.

### 2.6. Carbazole Alkaloids

*Murraya koenigii* (L.) *Spreng* is an endemic plant of India belonging to the *Rutaceae* family, also known as "meethi neem" or "curry patta" [42]. The extracts of this plant have been extensively investigated for their antidiabetic activity. Recent studies have demonstrated that the main phytocompounds of this plant are alkaloids whose structures contain a carbazole nucleus [42,72,73]. In 2010, Biswas et al. [39] investigated the hypoglycemic effects of mahanine (**11**) (Figure 9), a carbazole-containing alkaloid, isolated from the leaves of the plant. The results of the experiments on mice with diet-induced diabetes showed a significant reduction in hyperglycemia and insulin resistance, most likely due to the ability of this phytocompound to modulate the expression of the IR gene and the activation of the NF-jB pathway [74].

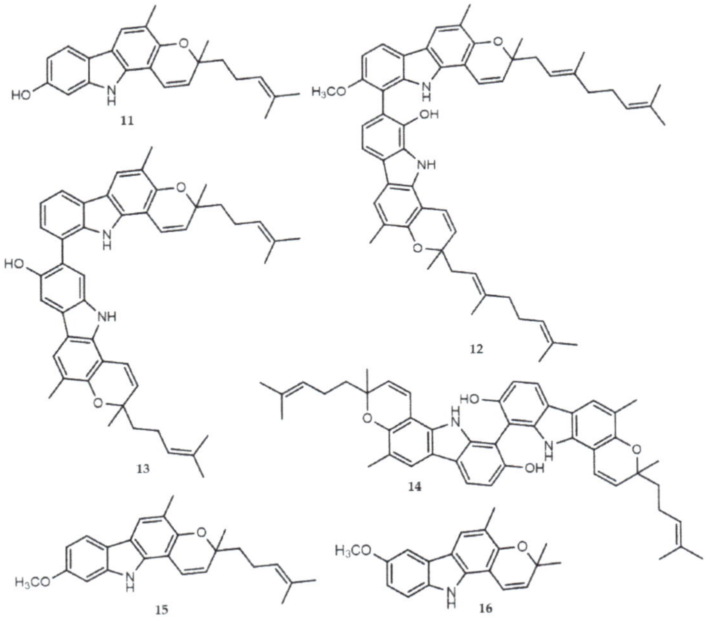

**Figure 9.** Structures of: mahanine (**11**), bisgerayafoline D (**12**), bismahanimbinol (**13**), bispyrayafoline (**14**), *O*-methyl mahanine (**15**) and *O*-methyl mukonal (**16**).

In a separate study, in addition to mahanine (**11**), Uvarani et al. isolated from the fruit pulp extract another six carbazole-containing alkaloids: bisgerayafoline D (**12**), bismahanimbinol (**13**), bispyrayafoline (**14**), *O*-methyl mahanine (**15**), *O*-methyl mukonal (**16**) (Figure 9). All these compounds (**11–16**) were tested for antioxidant, anti-α-glucosidase, DNA binding, protein interactions, and cytotoxic activities. Out of all the phytochemicals tested, mahanine (**11**) resulted to be the most promising as a radical scavenger and as an $\alpha$-glucosidase inhibitor ($IC_{50}$ of 21.4 $\pm$ 0.4 $\mu$M). Furthermore, this compound showed cytotoxic activity due to its ability to act as a DNA-intercalating agent [40].

The antidiabetic action of mahanine (**11**) was also confirmed in a study by Nooron et al. [41] in which the effects of the phytocompound on glucose uptake and translocation of glucose transporter 4 (GLUT4) in skeletal muscle and adipocyte cells were evaluated. In particular, the mahanine (**11**) treatment promoted a dose-dependent increase in glucose uptake in L6 myotubes and adipocyte cells through the stimulation of the Akt signaling pathway. These findings suggested that mahanine (**11**) acted similarly to insulin, promoting an enhancement of the Akt-mediated signaling pathway and leading to an increased translocation of GLUT4 on the cell membrane and, consequently, to a higher glucose uptake.

Carbazole-containing alkaloids (**16** (Figure 9) and **17–20** (Figure 10)) found in *M. koenigii* leaf extract promoted a substantial increase in glucose uptake, which was correlated to a higher translocation of GLUT4 into L6-GLUT4myc myotubes. The activity of compounds **16–19** was also investigated in rats with streptozotocin-induced diabetes. Compound **18**, koenidine, was found to be the most effective in lowering blood glucose and was therefore selected for further in vivo assays in leptin-receptor-deficient db/db mice. As expected, a significant increase in insulin sensitivity and a progressive lowering of blood glucose level were recorded.

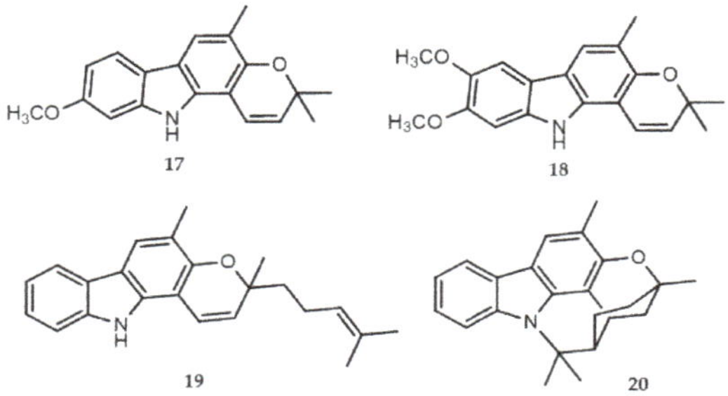

**Figure 10.** Structures of: O-methylmurrayamine A (**17**), koenidine (**18**), mahanimbine (**19**) and murrayazoline (**20**).

Furthermore, Western blot studies confirmed that the stimulation of GLUT4 translocation observed in L6-GLUT4myc myotubes is mediated by activation of the AKT-modulated signaling pathway. In vitro and in vivo pharmacokinetic studies revealed that compound **18** outperformed the other alkaloids due to its higher metabolic stability and systemic availability. The combination of properties of compound **18** makes it an ideal drug candidate for the treatment of diabetes and the treatment of insulin resistance [42].

### 2.7. Carbazole-Ethoxy-Phenyl Propionic Acid Derivative

The discovery of the nuclear receptor peroxisome proliferator-activated receptor (PPAR) -$\alpha$ and $\gamma$, as primary targets for fibrates and the thiazolidinediones (TZDs), respectively, has opened up new possibilities for the development of alternative tools for the treatment of type 2 DM [75,76].

The ability of TZD and fibrates to lower blood sugar and lipids, respectively, demonstrated in experimental models of insulin resistance and hyperlipidemia, has been known for decades. The improvement in insulin sensitivity promoted by TZDs (pioglitazone or rosiglitazone), although modest, is nevertheless significant and has justified the investments in the search for alternative diabetes treatments, using these compounds as starting compounds [77,78]. Thus far, several computational and in vitro studies on the PPAR receptor have been conducted for this purpose.

One of these research works has led to the identification of compound (*S*)-3-(4-(2-(9*H*-carbazol-9-yl)-ethoxy)phenyl)-2-ethoxypropanoic acid (**21**) (Figure 11), which showed dual activity on both PPAR isoforms $\alpha$ and $\gamma$ [43]. The results of an oral glucose tolerance test (OGTT) showed that treatment with compound **21** promoted an improvement in insulin sensitivity, greater than that seen with both pioglitazone and rosiglitazone. Compound **21** was also able to lower plasma concentrations of triglycerides and cholesterol in rats fed high cholesterol, whereas treatment with other PPAR$\gamma$ agonists did not have the same results [43].

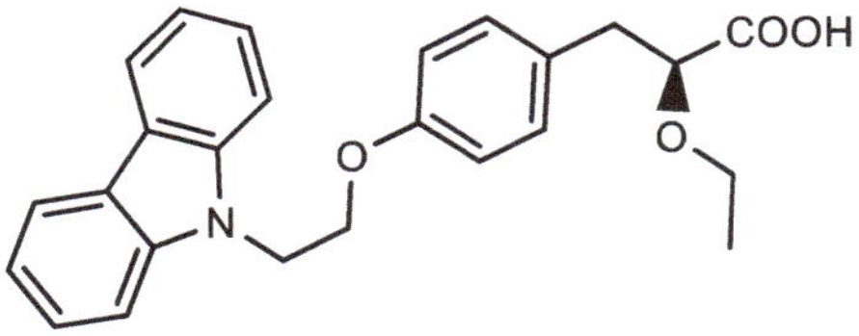

**Figure 11.** Structure of (*S*)-3-(4-(2-(9*H*-carbazol-9-yl)ethoxy)phenyl)-2-ethoxypropanoic acid (**21**).

*2.8. Chiglitazar*

Chiglitazar (Bilessglu®) (**22**) (Figure 12) is a small molecule with agonist activity against PPAR$\alpha$, $\delta$ and $\gamma$, recently developed by Chipscreen Bioscience.

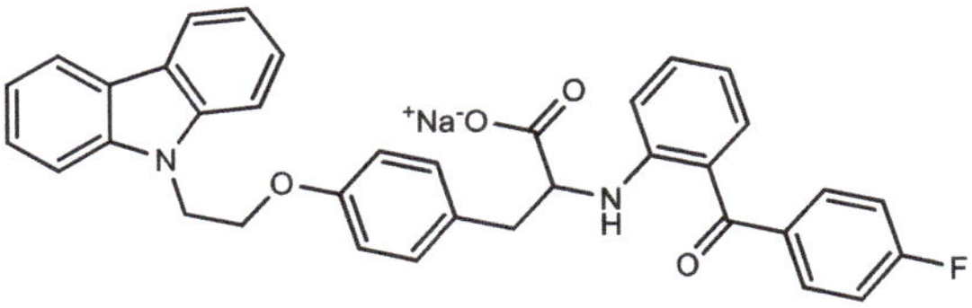

**Figure 12.** Structure of Chiglitazar (**22**).

In China, this drug was approved in October 2021 for the treatment of diabetes and is currently in phase II clinical trials for the treatment of non-alcoholic steatohepatitis [44]. In a very recent phase III study [45], the efficacy and safety of chiglitazar (**22**) were investigated in patients with type 2 DM who were unable to control their blood sugar with diet and exercise alone. Treatment with the drug resulted in significantly better blood glucose control than the placebo, with no significant toxic effects. Multiple administrations of different doses of chiglitazar (**22**) were generally well tolerated by patients of different ages.

An enhancement in insulin sensitivity and triglyceride level reduction were also recorded. Overall, clinical studies confirmed the drug's properties observed in preclinical experiments, bolstering the value of PPAR pan-agonists as efficient and safe tools for the treatment of diabetes [46].

## 3. Conclusions

In conclusion, we covered several studies on carbazoles (Table 2), both of synthetic and natural origin, having a role in the pathogenesis and development of diabetes. Some of these compounds, such as carvedilol (**1**) and mahanine (**11**), have been extensively investigated by different research groups, demonstrating a critical role in diabetes control due to their ability to modulate various pathways involved in the onset or evolution of the pathology.

Chiglitazar (**22**), one of the most promising carbazole derivatives, passed the preclinical experimentation and very recently entered clinical trials.

Important considerations also emerge from the structure–activity relationships of the described carbazole compounds. In particular, it seems that *N*-substitution with triazinic (see compound **4**) or triazolic portions (see compounds **7** and **8**) favors α-glucosidase inhibitory activity; the presence of cyclic sulfonamidic (see compound **5**) or cyclic urea groups (see compound **6**) modulates the cryptochrome activity, whereas the presence of ethylphenoxy groups improves insulin sensitivity (see compounds **21** and **22**). Moreover hydrogenation of a ring of the carbazole core favors the hypoglycemic effect via the AMPK pathway (see compounds **9** and **10**), and the tetracyclic system of the mahanine derivatives seems to be important in glucose uptake and translocation of glucose transporter GLUT4 in skeletal muscle and adipocyte cells (see compounds **11**, **16–20**). Finally compounds with a high conjugation effect show predominant antioxidant activity (see compounds **12–15**). Overall, the data herein reported could provide an important reference guide for the development of alternative and effective antidiabetic agents.

**Author Contributions:** Conceptualization, A.C.; methodology, F.G.; software, G.I.; resources, M.S.S. and C.S.; data curation, A.C. and G.I.; writing—original draft preparation, M.A.O.; writing—review and editing, F.G. and H.E.-K.; supervision, F.G., G.I. and A.C. All authors have read and agreed to the published version of the manuscript.

**Funding:** This research received no external funding.

**Institutional Review Board Statement:** Not applicable.

**Informed Consent Statement:** Not applicable.

**Data Availability Statement:** Not applicable.

**Conflicts of Interest:** The authors declare no conflict of interest.

## References

1. Kitabchi, A.E.; Umpierrez, G.E.; Miles, J.M.; Fisher, J.N. Hyperglycemic Crises in Adult Patients with Diabetes. *Diabetes Care* **2009**, *32*, 1335–1343. [CrossRef] [PubMed]
2. Baron, A.D. Postprandial Hyperglycaemia and Alpha-Glucosidase Inhibitors. *Diabetes Res Clin Pract.* **1998**, *40*, S51–S55. [CrossRef] [PubMed]
3. Lebovitz, H.E. Effect of the Postprandial State on Nontraditional Risk Factors. *Am. J. Cardiol.* **2001**, *88*, 5H–20H. [CrossRef]
4. Deshpande, A.D.; Harris-Hayes, M.; Schootman, M. Epidemiology of Diabetes and Diabetes-Related Complications. *Phys. Ther.* **2008**, *88*, 1254–1264. [CrossRef] [PubMed]
5. López-Candales, A. Metabolic Syndrome X: A Comprehensive Review of the Pathophysiology and Recommended Therapy. *J. Med.* **2001**, *32*, 283–300.
6. Wang, G.; Wang, J.; He, D.; Li, X.; Li, J.; Peng, Z. Synthesis and Biological Evaluation of Novel 1,2,4-Triazine Derivatives Bearing Carbazole Moiety as Potent α-Glucosidase Inhibitors. *Bioorg. Med. Chem. Lett.* **2016**, *26*, 2806–2809. [CrossRef] [PubMed]
7. Farahani-Zangaraki, M.; Taheri, A.; Etebari, M. Niosome-Carvedilol Protects DNA Damage of Supraphysiologic Concentrations of Insulin Using Comet Assay: An in Vitro Study. *Hum. Exp. Toxicol.* **2021**, *40*, S150–S157. [CrossRef]
8. Singh, K.; Maity, P.; Krug, L.; Meyer, P.; Treiber, N.; Lucas, T.; Basu, A.; Kochanek, S.; Wlaschek, M.; Geiger, H.; et al. Superoxide Anion Radicals Induce IGF-1 Resistance through Concomitant Activation of PTP 1 B and PTEN. *EMBO Mol. Med.* **2015**, *7*, 59–77. [CrossRef]
9. Rostoker, R.; Bitton-Worms, K.; Caspi, A.; Shen-Orr, Z.; LeRoith, D. Investigating New Therapeutic Strategies Targeting Hyperinsulinemia's Mitogenic Effects in a Female Mouse Breast Cancer Model. *Endocrinology* **2013**, *154*, 1701–1710. [CrossRef]
10. Kang, H.J.; Yi, Y.W.; Kim, H.J.; Hong, Y.B.; Seong, Y.S.; Bae, I. BRCA1 Negatively Regulates IGF-1 Expression through an Estrogen-Responsive Element-like Site. *Cell Death Dis.* **2012**, *3*, e336. [CrossRef]
11. Babiker, A.; Dubayee, M. Anti-Diabetic Medications: How to Make a Choice? *Sudan J. Paediatr.* **2017**, *17*, 11–20. [CrossRef]
12. Saturnino, C.; Grande, F.; Aquaro, S.; Caruso, A.; Iacopetta, D.; Bonomo, M.G.; Longo, P.; Schols, D.; Sinicropi, M.S. Chloro-1,4-Dimethyl-9H-Carbazole Derivatives Displaying Anti-HIV Activity. *Molecules* **2018**, *23*, 286. [CrossRef]
13. Ceramella, J.; Iacopetta, D.; Barbarossa, A.; Caruso, A.; Grande, F.; Bonomo, M.G.; Mariconda, A.; Longo, P.; Carmela, S.; Sinicropi, M.S. Carbazole Derivatives as Kinase-Targeting Inhibitors for Cancer Treatment. *Mini Rev. Med. Chem.* **2020**, *20*, 444–465. [CrossRef]
14. Grande, F.; de Bartolo, A.; Occhiuzzi, M.A.; Caruso, A.; Rocca, C.; Pasqua, T.; Carocci, A.; Rago, V.; Angelone, T.; Sinicropi, M.S. Carbazole and Simplified Derivatives: Novel Tools toward β-Adrenergic Receptors Targeting. *Appl. Sci.* **2021**, *11*, 5486. [CrossRef]
15. Caruso, A.; Voisin-Chiret, A.S.; Lancelot, J.C.; Sinicropi, M.S.; Garofalo, A.; Rault, S. Efficient and Simple Synthesis of 6-Aryl-1,4-Dimethyl-9H-Carbazoles. *Molecules* **2008**, *13*, 1312–1320. [CrossRef]
16. Chakraborty, D.P. Chapter 4 Chemistry and Biology of Carbazole Alkaloids. *Alkaloids Chem. Pharmacol.* **1993**, *44*, 257–364. [CrossRef]

17. Sinicropi, M.S.; Iacopetta, D.; Rosano, C.; Randino, R.; Caruso, A.; Saturnino, C.; Muià, N.; Ceramella, J.; Puoci, F.; Rodriquez, M.; et al. N-Thioalkylcarbazoles Derivatives as New Anti-Proliferative Agents: Synthesis, Characterisation and Molecular Mechanism Evaluation. *J. Enzym. Inhib. Med. Chem.* **2018**, *33*, 434–444. [CrossRef]

18. Bashir, M.; Bano, A.; Ijaz, A.S.; Chaudhary, B.A. Recent Developments and Biological Activities of N-Substituted Carbazole Derivatives: A Review. *Molecules* **2015**, *20*, 13496–13517. [CrossRef]

19. Caruso, A.; Chimento, A.; El-Kashef, H.; Lancelot, J.C.; Panno, A.; Pezzi, V.; Saturnino, C.; Sinicropi, M.S.; Sirianni, R.; Rault, S. Antiproliferative Activity of Some 1,4-Dimethylcarbazoles on Cells That Express Estrogen Receptors: Part, I.J. *Enzym. Inhib. Med. Chem.* **2012**, *27*, 609–613. [CrossRef]

20. Caruso, A.; Sinicropi, M.S.; Lancelot, J.C.; El-Kashef, H.; Saturnino, C.; Aubert, G.; Ballandonne, C.; Lesnard, A.; Cresteil, T.; Dallemagne, P.; et al. Synthesis and Evaluation of Cytotoxic Activities of New Guanidines Derived from Carbazoles. *Bioorg. Med. Chem. Lett.* **2014**, *24*, 467–472. [CrossRef]

21. Panno, A.; Sinicropi, M.S.; Caruso, A.; El-Kashef, H.; Lancelot, J.C.; Aubert, G.; Lesnard, A.; Cresteil, T.; Rault, S. New Trimethoxybenzamides and Trimethoxyphenylureas Derived from Dimethylcarbazole as Cytotoxic Agents. Part, I.J. *Heterocycl. Chem.* **2014**, *51*, E294–E302. [CrossRef]

22. Humphries, P.S.; Bersot, R.; Kincaid, J.; Mabery, E.; McCluskie, K.; Park, T.; Renner, T.; Riegler, E.; Steinfeld, T.; Turtle, E.D.; et al. Carbazole-Containing Amides and Ureas: Discovery of Cryptochrome Modulators as Antihyperglycemic Agents. *Bioorg. Med. Chem. Lett.* **2018**, *28*, 293–297. [CrossRef]

23. Sinicropi, M.; Lappano, R.; Caruso, A.; Santolla, M.; Pisano, A.; Rosano, C.; Capasso, A.; Panno, A.; Lancelot, J.; Rault, S.; et al. (6-Bromo-1,4-Dimethyl-9H-Carbazol-3-Yl-Methylene)-Hydrazine (Carbhydraz) Acts as a GPER Agonist in Breast Cancer Cells. *Curr. Top. Med. Chem.* **2015**, *15*, 1035–1042. [CrossRef] [PubMed]

24. Saturnino, C.; Caruso, A.; Iacopetta, D.; Rosano, C.; Ceramella, J.; Muià, N.; Mariconda, A.; Bonomo, M.G.; Ponassi, M.; Rosace, G.; et al. Inhibition of Human Topoisomerase II by N,N,N-Trimethylethanammonium Iodide Alkylcarbazole Derivatives. *ChemMedChem* **2018**, *13*, 2635–2643. [CrossRef] [PubMed]

25. Saturnino, C.; Sinicropi, M.S.; Iacopettab, D.; Ceramella, J.; Caruso, A.; Muià, N.; Longo, P.; Rosace, G.; Galletta, M.; Ielo, I.; et al. N-Thiocarbazole-Based Gold Nanoparticles: Synthesis, Characterization and Anti-Proliferative Activity Evaluation. *IOP Conf. Ser. Mater. Sci. Eng.* **2018**, *459*, 012023. [CrossRef]

26. Caruso, A.; Ceramella, J.; Iacopetta, D.; Saturnino, C.; Mauro, M.V.; Bruno, R.; Aquaro, S.; Sinicropi, M.S. Carbazole Derivatives as Antiviral Agents: An Overview. *Molecules* **2019**, *24*, 1912. [CrossRef]

27. Nodari, S.; Metra, M.; Dei Cas, A.; Dei Cas, L. Efficacy and Tolerability of the Long-Term Administration of Carvedilol in Patients with Chronic Heart Failure with and without Concomitant Diabetes Mellitus. *Eur. J. Heart Fail.* **2003**, *5*, 803–809. [CrossRef]

28. Kveiborg, B.; Hermann, T.S.; Major-Pedersen, A.; Christiansen, B.; Rask-Madsen, C.; Raunsø, J.; Køber, L.; Torp-Pedersen, C.; Dominguez, H. Metoprolol Compared to Carvedilol Deteriorates Insulin-Stimulated Endothelial Function in Patients with Type 2 Diabetes—A Randomized Study. *Cardiovasc Diabetol* **2010**, *9*, 21. [CrossRef]

29. Vardeny, O.; Nicholas, G.; Andrei, A.; Buhr, K.A.; Hermanson, M.P.; Moran, J.J.; Detry, M.A.; Stein, J.H. β-AR Polymorphisms and Glycemic and Lipid Parameters in Hypertensive Individuals Receiving Carvedilol or Metoprolol. *Am. J. Hypertens.* **2012**, *25*, 920–926. [CrossRef]

30. Nardotto, G.H.B.; Coelho, E.B.; Paiva, C.E.; Lanchote, V.L. Effects of Type 2 Diabetes Mellitus in Patients on Treatment with Glibenclamide and Metformin on Carvedilol Enantiomers Metabolism. *J. Clin. Pharm.* **2017**, *57*, 760–769. [CrossRef]

31. Nardotto, G.H.B.; Lanchote, V.L.; Coelho, E.B.; della Pasqua, O. Population Pharmacokinetics of Carvedilol Enantiomers and Their Metabolites in Healthy Subjects and Type-2 Diabetes Patients. *Eur. J. Pharm. Sci.* **2017**, *109*, S108–S115. [CrossRef] [PubMed]

32. Nguyen, L.V.; Ta, Q.V.; Dang, T.B.; Nguyen, P.H.; Nguyen, T.; Pham, T.V.H.; Nguyen, T.H.T.; Baker, S.; le Tran, T.; Yang, D.J.; et al. Carvedilol Improves Glucose Tolerance and Insulin Sensitivity in Treatment of Adrenergic Overdrive in High Fat Diet-Induced Obesity in Mice. *PLoS ONE* **2019**, *14*, e0224674. [CrossRef]

33. Amirshahrokhi, K.; Zohouri, A. Carvedilol Prevents Pancreatic β-Cell Damage and the Development of Type 1 Diabetes in Mice by the Inhibition of Proinflammatory Cytokines, NF-KB, COX-2, INOS and Oxidative Stress. *Cytokine* **2021**, *138*, 155394. [CrossRef] [PubMed]

34. Rahim, F.; Ullah, H.; Javid, M.T.; Wadood, A.; Taha, M.; Ashraf, M.; Shaukat, A.; Junaid, M.; Hussain, S.; Rehman, W.; et al. Synthesis, in vitro evaluation and molecular docking studies of thiazole derivatives as new inhibitors of α-glucosidase. *Bioorg. Chem.* **2015**, *62*, 15–21. [CrossRef] [PubMed]

35. Humphries, P.S.; Bersot, R.; Kincaid, J.; Mabery, E.; McCluskie, K.; Park, T.; Renner, T.; Riegler, E.; Steinfeld, T.; Turtle, E.D.; et al. Carbazole-Containing Sulfonamides and Sulfamides: Discovery of Cryptochrome Modulators as Antidiabetic Agents. *Bioorg. Med. Chem. Lett.* **2016**, *26*, 757–760. [CrossRef]

36. Iqbal, S.; Khan, M.A.; Javaid, K.; Sadiq, R.; Fazal-ur-Rehman, S.; Choudhary, M.I.; Basha, F.Z. New Carbazole Linked 1,2,3-Triazoles as Highly Potent Non-Sugar α-Glucosidase Inhibitors. *Bioorg. Chem.* **2017**, *74*, 72–81. [CrossRef]

37. Zhang, J.Q.; Li, S.M.; Ma, X.; Zhong, G.; Chen, R.; Li, X.S.; Zhu, G.F.; Zhou, B.; Guo, B.; Wu, H.S.; et al. Discovery of Tetrahydrocarbazoles with Potent Hypoglycemic and Hypolipemic Activities. *Eur. J. Med. Chem.* **2018**, *150*, 102–112. [CrossRef]

38. Wang, L.L.; Du, Y.; Li, S.M.; Cheng, F.; Zhang, N.N.; Chen, R.; Cui, X.; Yang, S.G.; Fan, L.L.; Wang, J.T.; et al. Design, Synthesis and Evaluation of Tetrahydrocarbazole Derivatives as Potential Hypoglycemic Agents. *Bioorg. Chem.* **2021**, *115*, 105172. [CrossRef]

39. Biswas, A.; Bhattacharya, S.; Dasgupta, S.; Kundu, R.; Roy, S.S.; Pal, B.C.; Bhattacharya, S. Insulin Resistance Due to Lipid-Induced Signaling Defects Could Be Prevented by Mahanine. *Mol. Cell Biochem.* **2010**, *336*, 97–107. [CrossRef]
40. Uvarani, C.; Jaivel, N.; Sankaran, M.; Chandraprakash, K.; Ata, A.; Mohan, P.S. Axially Chiral Biscarbazoles and Biological Evaluation of the Constituents from *Murraya Koenigii*. *Fitoterapia* **2014**, *94*, 10–20. [CrossRef]
41. Nooron, N.; Athipornchai, A.; Suksamrarn, A.; Chiabchalard, A. Mahanine Enhances the Glucose-Lowering Mechanisms in Skeletal Muscle and Adipocyte Cells. *Biochem. Biophys. Res. Commun.* **2017**, *494*, 101–106. [CrossRef]
42. Patel, O.P.S.; Mishra, A.; Maurya, R.; Saini, D.; Pandey, J.; Taneja, I.; Raju, K.S.R.; Kanojiya, S.; Shukla, S.K.; Srivastava, M.N.; et al. Naturally Occurring Carbazole Alkaloids from Murraya Koenigii as Potential Antidiabetic Agents. *J. Nat. Prod.* **2016**, *79*, 1276–1284. [CrossRef]
43. Sauerberg, P.; Pettersson, I.; Jeppesen, L.; Bury, P.S.; Mogensen, J.P.; Wassermann, K.; Brand, C.L.; Sturis, J.; Wöldike, H.F.; Fleckner, J.; et al. Novel Tricyclic-Alpha-Alkyloxyphenylpropionic Acids: Dual PPARalpha/Gamma Agonists with Hypolipidemic and Antidiabetic Activity. *J. Med. Chem.* **2002**, *45*, 789–804. [CrossRef]
44. Deeks, E.D. Chiglitazar: First Approval. *Drugs* **2022**, *82*, 87–92. [CrossRef]
45. Ji, L.; Song, W.; Fang, H.; Li, W.; Geng, J.; Wang, Y.; Guo, L.; Cai, H.; Yang, T.; Li, H.; et al. Efficacy and Safety of Chiglitazar, a Novel Peroxisome Proliferator-Activated Receptor Pan-Agonist, in Patients with Type 2 Diabetes: A Randomized, Double-Blind, Placebo-Controlled, Phase 3 Trial (CMAP). *Sci. Bull. Beijing* **2021**, *66*, 1571–1580. [CrossRef]
46. Li, X.; Yu, J.; Wu, M.; Li, Q.; Liu, J.; Zhang, H.; Zhu, X.; Li, C.; Zhang, J.; Ning, Z.; et al. Pharmacokinetics and Safety of Chiglitazar, a Peroxisome Proliferator-Activated Receptor Pan-Agonist, in Patients. *Clin. Pharm. Drug Dev.* **2021**, *10*, 789–796. [CrossRef]
47. Hercek, R.; Skoda, A.; Proksa, B. Process for Preparation of Carvedilol. U.S. Patent 2006/0167077A1, 19 May 2009.
48. Berridge, M.S.; Nelson, A.D.; Zheng, L.; Leisure, G.P.; Miraldi, F. Specific Beta-Adrenergic Receptor Binding of Carazolol Measured with PET. *J. Nucl. Med.* **1994**, *35*, 1665–1676.
49. Fonarow, G.C. Role of Carvedilol Controlled-Release in Cardiovascular Disease. *Expert. Rev. Cardiovasc. Ther.* **2009**, *7*, 483–498. [CrossRef]
50. Ahmad, A. Carvedilol Can Replace Insulin in the Treatment of Type 2 Diabetes Mellitus. *J. Diabetes Metab.* **2017**, *8*, 2. [CrossRef]
51. Bakris, G.L.; Fonseca, V.; Katholi, R.E.; McGill, J.B.; Messerli, F.H.; Phillips, R.A.; Raskin, P.; Wright, J.T.; Oakes, R.; Lukas, M.A.; et al. Metabolic Effects of Carvedilol vs Metoprolol in Patients with Type 2 Diabetes Mellitus and Hypertension: A Randomized Controlled Trial. *JAMA* **2004**, *292*, 2227–2236. [CrossRef]
52. Seferović, P.M.; Petrie, M.C.; Filippatos, G.S.; Anker, S.D.; Rosano, G.; Bauersachs, J.; Paulus, W.J.; Komajda, M.; Cosentino, F.; de Boer, R.A.; et al. Type 2 Diabetes Mellitus and Heart Failure: A Position Statement from the Heart Failure Association of the European Society of Cardiology. *Eur. J. Heart Fail.* **2018**, *20*, 853–872. [CrossRef]
53. Dézsi, C.A.; Szentes, V. The Real Role of B-Blockers in Daily Cardiovascular Therapy. *Am. J. Cardiovasc. Drugs* **2017**, *17*, 361–373. [CrossRef]
54. Jacob, S.; Rett, K.; Wicklmayr, M.; Agrawal, B.; Augustin, H.J.; Dietze, G.J. Differential Effect of Chronic Treatment with Two Beta-Blocking Agents on Insulin Sensitivity: The Carvedilol-Metoprolol Study. *J. Hypertens.* **1996**, *14*, 489–494. [CrossRef]
55. Refsgaard, J.; Thomsen, C.; Andreasen, F.; Gøtzsche, O. Carvedilol Does Not Alter the Insulin Sensitivity in Patients with Congestive Heart Failure. *Eur. J. Heart Fail.* **2002**, *4*, 445–453. [CrossRef]
56. Yue, T.L.; Cheng, H.Y.; Lysko, P.G.; McKenna, P.J.; Feuerstein, R.; Gu, J.L.; Lysko, K.A.; Davis, L.L.; Feuerstein, G. Carvedilol, a New Vasodilator and Beta Adrenoceptor Antagonist, Is an Antioxidant and Free Radical Scavenger. *J. Pharm. Exp. Ther.* **1992**, *263*, 92–98.
57. Afonso, R.A.; Patarrao, R.S.; Macedo, M.P.; Carmo, M.M. Carvedilol Action Is Dependent on Endogenous Production of Nitric Oxide. *Am. J. Hypertens.* **2006**, *19*, 419–425. [CrossRef]
58. Ehrenborg, E.; Skogsberg, J.; Ruotolo, G.; Large, V.; Eriksson, P.; Arner, P.; Hamsten, A. The Q/E27 Polymorphism in the Beta2-Adrenoceptor Gene Is Associated with Increased Body Weight and Dyslipoproteinaemia Involving Triglyceride-Rich Lipoproteins. *J. Intern. Med.* **2000**, *247*, 651–656. [CrossRef]
59. González Sánchez, J.L.; Proenza, A.M.; Martínez Larrad, M.T.; Ramis, J.M.; Fernández Pérez, C.; Palou, A.; Serrano Ríos, M. The Glutamine 27 Glutamic Acid Polymorphism of the Beta2-Adrenoceptor Gene Is Associated with Abdominal Obesity and Greater Risk of Impaired Glucose Tolerance in Men but Not in Women: A Population-Based Study in Spain. *Clin. Endocrinol. Oxf.* **2003**, *59*, 476–481. [CrossRef]
60. Suresha, R.N.; Ashwini, V.; Pragathi, B.; Kalabharathi, H.L.; Satish, A.M.; Pushpa, V.H.; Jayanthi, M.K.; Snehalatha, P. The Effect of Carvedilol on Blood Glucose Levels in Normal Albino Rats. *J. Clin. Diagn. Res.* **2013**, *7*, 1900–1903. [CrossRef]
61. Farhat, R.; Su, G.; Sejling, A.S.; Knight, N.; Fisher, S.J.; Chan, O. Carvedilol Prevents Counterregulatory Failure and Impaired Hypoglycaemia Awareness in Non-Diabetic Recurrently Hypoglycaemic Rats. *Diabetologia* **2019**, *62*, 676–686. [CrossRef]
62. Mishra, J.; Löbmann, K.; Grohganz, H.; Rades, T. Influence of Preparation Technique on Co-Amorphization of Carvedilol with Acidic Amino Acids. *Int. J. Pharm.* **2018**, *552*, 407–413. [CrossRef]
63. Ross, S.A.; Gulve, E.A.; Wang, M. Chemistry and Biochemistry of Type 2 Diabetes. *Chem. Rev.* **2004**, *104*, 1255–1282. [CrossRef]
64. Bass, J. Circadian Topology of Metabolism. *Nature* **2012**, *491*, 348–356. [CrossRef]
65. Marcheva, B.; Ramsey, K.M.; Peek, C.B.; Affinati, A.; Maury, E.; Bass, J. Circadian Clocks and Metabolism. *Handb. Exp. Pharm.* **2013**, *217*, 127–155. [CrossRef]

66.  Renström, F.; Koivula, R.W.; Varga, T.V.; Hallmans, G.; Mulder, H.; Florez, J.C.; Hu, F.B.; Franks, P.W. Season-Dependent Associations of Circadian Rhythm-Regulating Loci (CRY1, CRY2 and MTNR1B) and Glucose Homeostasis: The GLACIER Study. *Diabetologia* **2015**, *58*, 997–1005. [CrossRef]

67.  Machicao, F.; Peter, A.; Machann, J.; Konigsrainer, I.; Bohm, A.; Lutz, S.Z.; Heni, M.; Fritsche, A.; Schick, F.; Konigsrainer, A.; et al. Glucose-Raising Polymorphisms in the Human Clock Gene Cryptochrome 2 (CRY2) Affect Hepatic Lipid Content. *PLoS ONE* **2016**, *11*, e0145563. [CrossRef]

68.  Kelly, M.A.; Rees, S.D.; Hydriem, Z.L.; Shera, A.S.; Bellary, S.; O'Hare, J.P.; Kumar, S.; Taheri, S.; Basit, A.; Barnett, A.H. Circadian Gene Variants and Susceptibility to Type 2 Diabetes: A Pilot Study. *PLoS ONE* **2012**, *7*, e32670. [CrossRef]

69.  Dashti, H.S.; Smith, C.E.; Lee, Y.C.; Parnell, L.D.; Lai, C.Q.; Arnett, D.K.; Ordovás, J.M.; Garaulet, M. CRY1 Circadian Gene Variant Interacts with Carbohydrate Intake for Insulin Resistance in Two Independent Populations: Mediterranean and North American. *Chronobiol. Int.* **2014**, *31*, 660–667. [CrossRef]

70.  Hardie, D.G. AMP-Activated Protein Kinase: Maintaining Energy Homeostasis at the Cellular and Whole-Body Levels. *Annu. Rev. Nutr.* **2014**, *34*, 31–55. [CrossRef]

71.  Herzig, S.; Shaw, R.J. AMPK: Guardian of Metabolism and Mitochondrial Homeostasis. *Nat. Rev. Mol. Cell Biol.* **2018**, *19*, 121–135. [CrossRef]

72.  Caruso, A.; Barbarossa, A.; Carocci, A.; Salzano, G.; Sinicropi, M.S.; Saturnino, C. Carbazole Derivatives as STAT Inhibitors: An Overview. *Appl. Sci.* **2021**, *11*, 6192. [CrossRef]

73.  Caruso, A.; Iacopetta, D.; Puoci, F.; Rita Cappello, A.; Saturnino, C.; Stefania Sinicropi, M. Carbazole Derivatives: A Promising Scenario for Breast Cancer Treatment. *Mini-Rev. Med. Chem.* **2016**, *16*, 630–643. [CrossRef] [PubMed]

74.  Santomauro, A.T.M.G.; Boden, G.; Silva, M.E.R.; Rocha, D.M.; Santos, R.F.; Ursich, M.J.M.; Strassmann, P.G.; Wajchenberg, B.L. Overnight Lowering of Free Fatty Acids with Acipimox Improves Insulin Resistance and Glucose Tolerance in Obese Diabetic and Nondiabetic Subjects. *Diabetes* **1999**, *48*, 1836–1841. [CrossRef] [PubMed]

75.  Willson, T.M.; Cobb, J.E.; Cowan, D.J.; Wiethe, R.W.; Correa, I.D.; Prakash, S.R.; Beck, K.D.; Moore, L.B.; Kliewer, S.A.; Lehmann, J.M. The Structure-Activity Relationship between Peroxisome Proliferator-Activated Receptor Gamma Agonism and the Antihyperglycemic Activity of Thiazolidinediones. *J. Med. Chem.* **1996**, *39*, 665–668. [CrossRef] [PubMed]

76.  Brown, P.J.; Winegar, D.A.; Plunket, K.D.; Moore, L.B.; Lewis, M.C.; Wilson, J.G.; Sundseth, S.S.; Koble, C.S.; Wu, Z.; Chapman, J.M.; et al. A Ureido-Thioisobutyric Acid (GW9578) Is a Subtype-Selective PPARalpha Agonist with Potent Lipid-Lowering Activity. *J. Med. Chem.* **1999**, *42*, 3785–3788. [CrossRef] [PubMed]

77.  Collins, J.L.; Blanchard, S.G.; Boswell, G.E.; Charifson, P.S.; Cobb, J.E.; Henke, B.R.; Hull-Ryde, E.A.; Kazmierski, W.M.; Lake, D.H.; Leesnitzer, L.M.; et al. N-(2-Benzoylphenyl)-L-Tyrosine PPARgamma Agonists. 2. Structure-Activity Relationship and Optimization of the Phenyl Alkyl Ether Moiety. *J. Med. Chem.* **1998**, *41*, 5037–5054. [CrossRef] [PubMed]

78.  Henke, B.R.; Blanchard, S.G.; Brackeen, M.F.; Brown, K.K.; Cobb, J.E.; Collins, J.L.; Harrington, W.W.; Hashim, M.A.; Hull-Ryde, E.A.; Kaldor, I.; et al. N-(2-Benzoylphenyl)-L-Tyrosine PPARgamma Agonists. 1. Discovery of a Novel Series of Potent Antihyperglycemic and Antihyperlipidemic Agents. *J. Med. Chem.* **1998**, *41*, 5020–5036. [CrossRef]

*Review*

# An Update of Carbazole Treatment Strategies for COVID-19 Infection

Maria Grazia Bonomo [1,†], Anna Caruso [1,2,*,†], Hussein El-Kashef [3,4], Giovanni Salzano [1], Maria Stefania Sinicropi [2,‡] and Carmela Saturnino [1,‡]

[1] Department of Science, University of Basilicata, 85100 Potenza, Italy
[2] Department of Pharmacy, Health and Nutritional Sciences, University of Calabria, 87036 Arcavacata Cosenza, Italy
[3] Faculty of Science, Assiut University, Assiut 71516, Egypt
[4] Faculty of Pharmacy, Sphinx University, New Assiut 71515, Egypt
[*] Correspondence: anna.caruso@unical.it or anna.caruso@unibas.it
[†] These authors contributed equally to this work.
[‡] Senior co-authors.

**Abstract:** The Coronavirus disease 2019 (COVID-19) outbreak was declared by the World Health Organization (WHO) in March 2020 to be a pandemic and many drugs used at the beginning proved useless in fighting the infection. Lately, there has been approval of some new generation drugs for the clinical treatment of severe or critical COVID-19 infections. Nevertheless, more drugs are required to reduce the pandemic's impact. Several treatment approaches for COVID-19 were employed since the beginning of the pandemic, such as immunomodulatory, antiviral, anti-inflammatory, antimicrobial agents, and again corticosteroids, angiotensin II receptor blockers, and bradykinin B2 receptor antagonists, but many of them were proven ineffective in targeting the virus. So, the identification of drugs to be used effectively for treatment of COVID-19 is strongly needed. It is aimed in this review to collect the information so far known about the COVID-19 studies and treatments. Moreover, the observations reported in this review about carbazoles as a treatment can signify a potentially useful clinical application; various drugs that can be introduced into the therapeutic equipment to fight COVID-19 or their molecules can be used as the basis for designing new antivirals.

**Keywords:** COVID-19; coronaviruses; SARS-CoV-2; carbazoles; alkaloids

**Citation:** Bonomo, M.G.; Caruso, A.; El-Kashef, H.; Salzano, G.; Sinicropi, M.S.; Saturnino, C. An Update of Carbazole Treatment Strategies for COVID-19 Infection. *Appl. Sci.* **2023**, *13*, 1522. https://doi.org/10.3390/app13031522

Academic Editor: Snezana Agatonovic-Kustrin

Received: 20 November 2022
Revised: 11 January 2023
Accepted: 21 January 2023
Published: 24 January 2023

## 1. Introduction

Coronaviruses (CoV) are a wide respiratory class of viruses that can induce mild to moderate diseases, ranging from the common cold to respiratory syndromes, such as Middle East Respiratory Syndrome (MERS), Severe Acute Respiratory Syndrome (SARS), and recently Severe Acute Respiratory Syndrome Coronavirus-2 (SARS-CoV-2) [1,2]. Coronaviruses are widespread in species of animals (such as camels and bats), but in different cases they can infect humans and then spread over among people. To date, seven types of human coronaviruses are known and are common around the world. The former ones were identified in the mid-1960s, others were only discovered in the new millennium. SARS-CoV-2 belongs to the family of Coronavirus discovered in humans and it's the cause of the worldwide human respiratory disease pandemic coronavirus 2019 (COVID-19), owing to its high ability to inter-transfer with humans [3]. The COVID-19 pandemic was discovered initially in Wuhan City, Central China, and it quickly spread over other countries and became a threat to the world health as it affected millions of people [4]; to date, COVID-19 represents a critical threat to global health and economy as more than 520 million people were infected and about 6 million died [5].

The pandemic outbreak has reached alarming proportions, blocking and putting pressure on national health systems and requiring a significant deployment of forces and resources around the world. The WHO declared it a pandemic on 11 March 2020 because of its very rapid transmission and the considerable rate of mortality and morbidity [6,7]. The majority of the infected people (80%) are asymptomatic or have non-severe symptoms, nevertheless, 15–20% of other patients need hospitalization as a result of the development of acute lung injury (ALI) and/or distress syndrome respiratory (ARDS) [8,9].

Recently, COVID-19 has been considered a multi-organ disease characterized by a wide spectrum of manifestations. Early variants of COVID-19 [10] cause a rise in body temperature, cough, fatigue and myalgia, and bacterial superinfection in most patients infected with SARS-CoV-2. Symptoms became worse in more severe infections, leading to the development of COVID-19-associated acute respiratory distress syndrome (CARDS) with respiratory failure. In such a case, patients should be hospitalized in an intensive care unit (ICU). Currently, COVID-19 has become very flu-like in most cases, but we don't know what will happen next; so, even though the COVID-19 era is not over yet [7], we have entered now the era of post-COVID-19. Many people who recovered from COVID-19 present a condition called "post-COVID syndrome" [11], in which they develop new or persisting symptoms that continue for a long time; this syndrome has been termed a second pandemic and has gradually been accepted as a new clinical problem to be addressed in relation with SARS-CoV-2 infection [12]. The increasing number of patients suffering from post-SARS-CoV-2 infection have numerous symptoms, such as neurocognitive, autonomic, gastrointestinal, respiratory, musculoskeletal, psychological, and other symptoms and manifestations related to post-COVID [7,13].

## 2. Viral Features

SARS-CoV-2 is a single-stranded RNA+ virus that belongs to the *Coronaviridae* family. It is broadly spread among humans and other mammals. The Severe Acute Respiratory Syndrome Coronavirus-2 possesses a broad genome (around 29,700 nucleotides) encoding structural and non-structural proteins [14]. The replication and transcription processes of the SARS-CoV-2 virus occur at the level of the cytoplasmic membrane by the action of viral replicase; the replicase gene encodes two overlapping polyproteins, pp1a and pp1ab, which allow viral replication and transcription [15]. These polyproteins are digested by the 33.8 kDa Mpro protease activity in at least 11 conserved sites. Mpro is known as an interesting target for the design of antiviral drugs because of the functional importance that Mpro has in the virus life cycle and because of the absence of closely related homologs in humans [16]. Other proteins have also been identified for their implication in the pathogenesis of the virus, such as the spike glycoprotein, which allows virus entry by binding to the angiotensin 2 converting enzyme receptor (ACE2) widely expressed in human tissues [5]. Possible drug targets include RNA polymerase-dependent RNA and spike (S) proteins [17] that bind to ACE2 and allow the virus to enter cells. Consequently, the virus's RNA polymerase enzyme is primed for the rapid mutation required to resist current antiviral drugs [2].

## 3. Different COVID-19 Treatments

In March 2020, he WHO declared the outbreak of COVID-19 as a pandemic, and several drugs used at the beginning proved useless in fighting the infection [6,7]. Lately, there has been the approval of some new-generation drugs for the clinical treatment of severe and critical COVID-19 symptoms. Nevertheless, more medications are required to attenuate the pandemic impact [5]. Many of the employed therapeutic approaches since the start of the pandemic were shown to be ineffective against the virus. So, the identification of drugs to be used effectively for the treatment of COVID-19 is strongly needed [18–20]. COVID-19 is considered to be a multisystem disease, and its pathogenesis involves severe lung inflammation and immune deficiency linked to inadequate immune response and enhanced cytokine production. Therefore, therapeutic approaches currently being studied concern antiviral and anti-proinflammatory cytokines, anti-infective and life support treatments, monoclonal antibodies, and passive immunotherapy, mostly in people with severe disease [21]. The antiviral treatments used so far for SARS-CoV-2 virus can be presented into two categories [22,23]:

(A)  Agents that target the virus proteins: S protein and viral proteases or virus RNA which are the main viral targets.

(B)  Agents that target host proteins that allow the virus to enter the cell, such as the enzyme ACE-2, TMPRSS2 (transmembrane protease serine 2), furin, and cathepsin-L, the proteins which can promote the attachment of viral cells such as HSPG (heparin sulfate proteoglycans), eukaryotic translation proteins such as S1R (endoplasmic reticulum chaperone protein), and transcriptional system proteins such as inosine monophosphate dehydrogenase and dihydroorotate dehydrogenase [22,23].

Several studies and laboratory works are still in demand to discover new and effective antiviral drugs against SARS-CoV-2. These studies could deal with the following areas [23]:

(1)  Inhibitors that block the virus from entering the human cell:

    a.  S Protein inhibitors: plasma from convalescing infected persons, miniproteins, monoclonal antibodies, nanocods, soluble human ACE-2 protein;

    b.  Fusion entry inhibitors;

    c.  inhibitors of TMPRSS2, such as nitazoxanid, camostat, nafamostat, gabexat, dutasteride, bromhexine, niclosamide, and proxalutamide;

    d.  Endosomal entry inhibitors: NIP1 inhibitors (EG00229), niclosamide, hydroxychloroquine, umifenovir, nitazoxanide; furin inhibitors (dec-RVKR-cmk); cathepsin L inhibitors (teicoplanin, SSAA09E1, K1777);

    e.  Inhibitors of HSPG (lactoferrin).

(2)  Viral proteases inhibitors: inhibitors of Mpro, the main protease of the virus (lopinavir/ritonavir, PF-07321332, PF-07304814, GC376, carprofen (**1**, Figure 1)); viral papain-like protease inhibitors (PLpro).

(3)  Viral RNA inhibitors, RNA-dependent RNA polymerase (RdRp) inhibitors (AT-527, remdesivir, molnupiravir, favipiravir), host protein inhibitors that support the synthesis of viral RNA (dihydroorarate dehydrogenase inhibitor (PTC299), inosine monophosphate dehydrogenase inhibitor (merimepodib).

(4)  Host protein inhibitors that support the synthesis of viral protein: S1R agonists (fluvoxamine); inhibitors eEF1A (plitidepsin).

(5)  Viral immunomodulation inhibitors: host $\alpha/\beta$ importin inhibitors (ivermectin).

(6)  Agents that support natural host immunity: interferons [23].

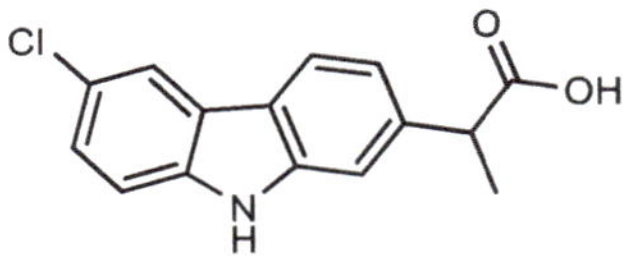

**Figure 1.** Structure of carprofen (**1**).

Previous studies to develop anti-virus SARS-CoV-2 therapies relied on the design of medicaments that obstruct the viral life cycle, so: (1) the inhibition of viral entry into the cell by blocking the spike protein and/or ACE2; (2) the inhibition of viral proteases; (3) the prevention of viral replication via the RNA dependent viral RNA polymerase (RdRp) inhibition [24]. Much of this research is based on the medicaments reuse strategy to find new therapies from already adopted drugs, speeding up like this the drug discovery process and providing the benefit of instant use of medicaments that have previously proven safety safe [25]. De novo drug identification is a long and costly process, so several scientists around the world have tested the action of already approved drugs on SARS-CoV-2. Lately, the FDA has agreed to a combination of lately developed antiviral drugs to treat severe or critical cases. However, the impact of this pandemic can be reduced by using other drugs [5]. It has been shown that emodin, omipalisib, and tipifarnib possess an inhibitory effect on RdRp [26], while in other studies it was shown that raltitrexed, cytarabin, lamivudine, tenofovir, cidofovir, and fludarabine are capable to bind the viral spike protein [27,28]. In clinical trials, antivirals such as ritonavir and ribavirin, have been utilized to treat COVID-19 [29–31]. The efficacious treatment development for this disease has been hindered by the limited data available on this coronavirus during the pandemic's start. The repurposing of drugs allows exploration of the possible antiviral activity of approved drugs. Furthermore, pharmacological synergism has better efficacy and reduced toxicity in antiviral treatment [5]. In an interesting study by Abdel-Halim et al. [5] an *in-silico* method has been proposed to observe the effect of approved drugs on an essential target of SARS-CoV-2, the principal protease (Mpro), to search rapid antiviral treatments and/or drug combinations, because of the reduced time and urgency of treatment; the inhibiting activity of three pharmaceutical compounds and two drug combinations on this protease was established. It has been proven, by *in-silico* and in-vitro assays, that favipiravir, carvedilol (**2**, Figure 2), and cefixime are Mpro inhibitors. The various conformational changes of the enzyme that can be induced by the binding of ligands were investigated using molecular dynamics simulations that identified potential drug combinations that show a synergistic effect when tested on Mpro; two drug combinations (favipiravir/cefixime and favipiravir/carvedilol (**2**, Figure 2)) have been shown to be active against this enzyme [5]. The study of Ahmad et al. [2] developed a potential strategy for the use of plant extracts, and in particular active phytochemicals from *Daphne* species plants, to tackle the SARS-CoV-2 pandemic. The studied compounds showed good antioxidant activity, which could be interesting if evaluated as radical scavenging capacity. Phytochemicals analyzed were found to be able inhibitors of SARSCoV-2 as they prevent the development of viral proteins along with the spread of infection, inhibiting the viral protein, leading the virus to lose the 6LU7 protein, the main viral protease. The synergic interactions of twelve coumarins showed good inhibition of the 6LU7 viral protein [2].

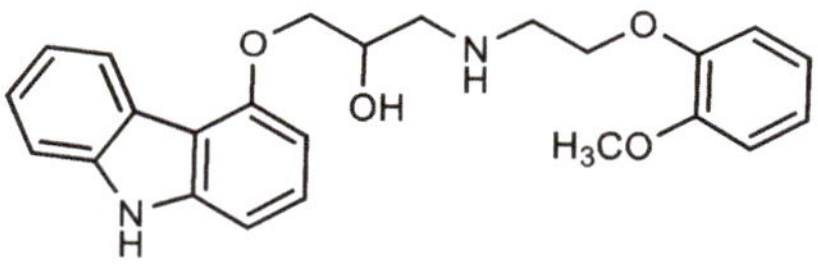

**Figure 2.** Structure of carvedilol (**2**).

Therefore, conventional treatments for COVID-19 include the use of anti-inflammatory, antiviral, and antimicrobial drugs, along with the use of immunomodulators, bradykinin B2 receptor antagonists, angiotensin II receptor blockers, and corticosteroids. Furthermore, an advantageous treatment to prevent or treat COVID-19 disease or post-COVID-19 syndrome is represented by the use of nutraceuticals [32–36]. For the COVID-19 treatment, the nutraceuticals' use was assessed by their interaction with the ACE2 enzyme. When the spike glycoprotein of the SARS-CoV-2 virus binds to the ACE2 receptor, there is the downregulation of ACE2 and the consequent improvement of the level of angiotensin-2 (Ang II) and the increase of the activation of the type 1 axis (AT1R) of the Ang II/Ang II receptor which is correlated with pro-inflammatory responses. Consequently, natural compounds play an important role in the treatment of COVID-19 because they decrease the activity of ACE2; the SARS-CoV-2 virus uses spike glycoprotein to enter host cells because the binding domain of the spike glycoprotein receptor (RBD) interacts with ACE2 on host cells; in the omicron variant of the virus, spike glycoprotein is characterized by 32 mutations, of which 15 are in RBD [37,38]. However, when there is a lack of effective curative and prophylactic drugs or when several mutants of the SARS-CoV-2 virus develop and spread rapidly among people, a key defense is a strong immune system [7], as the weakened immune system along with the older age, obesity and other diseases are risk factors that increase the severity of COVID-19 disease, and therefore supplements, nutraceuticals, and probiotics can reduce the risk of SARS-CoV-2 infection or alleviate COVID-19 symptoms [39,40]. Other studies highlighted the role that some bacterial substances and molecular compounds can play in the immune response to respiratory viruses and in the regulation of two main COVID-19 disease manifestations, which are systemic inflammation and endothelial damage. Additionally, the use of prebiotics, probiotics, and postbiotics has also been investigated in the battle against SARS-CoV-2 infection because the lack of these nutrients could result in a better susceptibility to infections and immune system dysfunctions. A clinical study demonstrated that the progression of COVID-19 can be regulated by the use of these supplements that modify the gut microbiota with a reduction of disease course and the severity of the symptoms [41,42]. The high dose administration of vitamin D, vitamin C, vitamin E, flavonoids, zinc, probiotics, prebiotics, omega-3 fatty acids, and melatonin are being used as principal dietary supplements in COVID-19 and presented a clinical benefit for the hospitalized; the viral load, the severity of the disease, and, therefore, the hospital stay are reduced by the immunomodulatory and antioxidant effects of these supplements [7,43].

It is of interest to note that the U.S. Food and Drug Administration (FDA) has approved the antiviral drug Remdesivir (Veklury) for the treatment of coronavirus disease 2019. This drug was prescribed as a treatment for hospitalized patients who are in need of supplemental oxygen or are at higher risk. This drug blocks the specific enzyme activity needed for the COVID-19 virus to replicate. Another approved medicament is Paxlovid. This medicament is used to treat patients with mild to moderate infections who are at higher risk of serious illness. Paxlovid is composed of two medicaments: nirmatrelvir and ritonavir. Nirmatrelvir blocks the activity of a specific enzyme needed for the virus to replicate, while the antiviral ritonavir helps to slow the breakdown of nirmatrelvir. Also, another drug named molnupiravir was approved for the treatment of mild to moderate infections in patients who are at higher risk of serious illness and are not able to take other medications. The anti-rheumatoid arthritis drug baricitinib (Olumiant) possesses also antiviral activity and it likely works against the virus by reducing the inflammation. The anti-inflammatory therapy, using anti-inflammatory drugs such as the corticosteroid dexamethasone, can treat or prevent organ dysfunction and lung injury from inflammation [44].

## 4. Carbazoles Treatment

Several studies [5,9,17,45–62] describing the important role of carbazole derivatives in the treatment of COVID-19 are included in this review, cf. Table 1.

**Table 1.** Carbazole derivatives-based drugs for the treatment of COVID-19.

| Compound | Name | Reference |
|---|---|---|
| SARS-CoV-2 M-pro inhibitors | | |
| 1 | Carprofen<br>2-(6-Chloro-9*H*-carbazol-2yl)propanoic acid | [45] |
| 2 | Carvedilol<br>1-(9*H*-Carbazol-4-yloxy)-3-[2-(2-methoxyphenoxy)ethylamino]propan-2-ol | [5,9,17,46–52] |
| 3 | Koenigicine<br>8-Methoxy-3,3,5-trimethyl-11*H*-pyrano[3,2-*a*]carbazol-9-ol | [53] |
| 4 | Mukonicine<br>9,11-Dimethoxy-3,3,5-trimethyl-11*H*-pyrano[3,2-*a*]carbazole | [53] |
| 5 | O-methylmurrayamine A<br>9-Methoxy-3,3,5-trimethyl-11*H*-pyrano[3,2-*a*]carbazole | [53] |
| 6 | Koenine<br>3,3,5-Trimethyl-11*H*-pyrano[3,2-*a*]carbazol-8-ol | [53] |
| 7 | Girinimbine<br>3,11-Dihydro-3,3,5-trimethyl-pyrano[3,2-*a*]carbazole | [53] |

**Table 1.** *Cont.*

| Compound | Name | Reference |
|---|---|---|
| SARS-CoV-2 M-pro inhibitors | | |
| Viral-entry inhibitors targeting human ACE2 | | |
| 8 | Edotecarin<br>6-((1,3-Dihydroxypropan-2-yl)amino)-2,10-dihydroxy-12-((2R,3R,4S,5S,6R)-3,4,5-trihydroxy-6-(hydroxymethyl)tetrahydro-2H-pyran-2-yl)-12,13-dihydro-5H-indolo[2,3-a]pyrrolo[3,4-c]carbazole-5,7(6H)-dione | [54,55] |
| 9 | 7-Hydroxystaurosporine<br>(5S,6R,7R,9R,16R)-16-hydroxy-6-methoxy-5-methyl-7-(methylamino)-6,7,8,9,15,16-hexahydro-17-oxa-4b,9a,15-triaza-5,9-methanodibenzo[b,h]cyclonona[jkl]cyclopenta[e]-as-indacen-14(5H)-one | [56] |
| 10 | CIMSSNa<br>sodium 3-(4-(((S)-5-((5S,7S,8R,9S)-8-methoxy-9-methyl-16-oxo-6,7,8,9,15,16-hexahydro-5H,14H-4b,9a,15-triaza-5,9-methanodibenzo[b,h]cyclonona[jkl]cyclopenta[e]-as-indacen-7-yl)-4-oxohexanamido)methyl)-1H-1,2,3-triazol-1-yl)propane-1-sulfonate | [57] |
| 11 | 6-Formylindolo(3,2-b)carbazole | [58] |
| NPC1 inhibitor | | |
| 12 | 2-((2-(1-Benzylpiperidin-4-yl)ethyl)amino)-N-(9H-carbazol-9-yl)acetamide | [59] |
| Antiviral against PLpro | | |
| 13 | 6-Cyano-5-methoxy-12-methylindolo [2, 3A] carbazole | [60] |
| Immunotherapy treatment | | |
| 14 | Ramatroban<br>3-[(3R)-3-[(4-fluorophenyl)sulfonylamino]-1,2,3,4-tetrahydrocarbazol-9-yl]propanoic<br>acid | [53,61,62] |

Carbazole containing heterocycles can be of synthetic or natural origin, many are alkaloids extracted from the leaves of *Ochrosia Elliptica Labill* [63]. Carbazoles represent an interesting class of heterocycles known by their anticancer activity [64,65]: antibacterial, anti-inflammatory [66], antifungal, antioxidant, antimicrobial, antiepileptic, antihistamine, antiviral [67,68]. In addition, numerous carbazole derivatives have also been found to be useful for Alzheimer's disease [69]. Several carbazoles were assessed by research groups for a SARS-CoV-2 virus study [5,45,53,54,56–60,70]. Despite vaccination against COVID-19, there is an urgent priority to identify additional antiviral drugs due to immune loss due to new variants of SARS-CoV-2; computational approaches have made a big contribution to the identification of antivirals by lowering costs and time and by accelerating analyzes of target interactions with candidate drugs [45,71]. A therapeutic strategy for drug development is the targeting of the main protease (Mpro) for its important role in the viral replication cycle. The overlapping polyproteins (pp1a and pp1ab) are cleaved, in an autoproteolytic way, by SARS-CoV-2 Mpro (SC2-Mpro), and mature non-structural proteins (11 proteins) necessary for viral replication and transcription are produced [72]. The lack of a Mpro-like human protease, such as its unique selective ability of cleavage site, makes it a major therapeutic target for the detection of anti-SARS-CoV-2 drugs [53,70]. Moreover, ACE2 is used as a therapeutic target to control the COVID-19 outbreak, as an ACE2-mediated mechanism allows the virus to enter permissive cells. ACE2 has a defensive role in heart disease as it is a membrane-bound zinc metallopeptidase that produces the vasodilator peptide angiotensin 1e7 [54]. Additionally, in order to identify new anti-COVID-19 drugs, an interaction between the SARS-CoV-2 nucleoprotein (N) and the cholesterol transporter Niemann-Pick type C1 (NPC1) [59] was evaluated. The NPC1 receptor, an endosomal membrane protein, regulates intracellular cholesterol transport. Another important protein in coronavirus is PLpro (papain-like protease), which performs an essential action in the processing mechanism of viral polyproteins [73] and has fundamental action against human immunity through post-translational modifications on human proteins [60,74]. Below, we report the drugs with a carbazole structure (**1–14**) that have been studied and selected to date, which can be inserted into the therapeutic equipment of the battle against COVID-19, or their scaffolds can be used as skeletons for the design of new antiviral compounds.

### 4.1. SARS-CoV-2 M-Pro Inhibitors

As already reported, protease M-pro, also known as chymotrypsin-like protease, (3CL-pro) is an enzyme that only exists in the virus and not in humans [75,76]. For this reason, Mpro is an interesting target for the discovery of new antivirals [77]. Gimeno et al. in 2020 [45] applied a virtual screening (VS) method for checking approved medicines to verify which of them could inhibit this protease. The drugs studied were docked against the structure of the protease involved using docking programs such as Glide, FRED, and AutoDock Vina. Thanks to these studies, drugs were selected, including one with a carbazole structure, carprofen (**1**, Figure 1), as possible M-pro inhibitors. Carprofen (2-(6-chloro-9*H*-carbazol-2yl) propanoic acid) (**1**) is a selective COX-2 (cyclooxygenase-2) inhibitor. Compound **1**, in the active site of M-pro, makes many interactions, such as hydrophobic interactions, with Gln189 and Met49, π-π interaction with His41 through its ring system, hydrogen interaction, for example with His164, Ser144, and Cys145, and halogen bond interaction with the thiol group of Cys44 through its chloro group. In-vitro studies show a limited inhibitory capacity on M-pro (3.97% at 50 μM) of drug **1** therefore this molecule could be considered a promising compound for the synthesis of more potent and effective inhibitors [45].

In 2022, Abdel-Halim et al. [5] identified carvedilol ((1-(9*H*-carbazol-4-yloxy)-3-[2-(2-methoxyphenoxy)ethylamino]propan-2-ol)) (**2**, Figure 2), a β-adrenergic receptor blocker, as Mpro inhibitor by *in-silico* and in-vitro assays.

In silico studies allowed us to study the conformational changes in the enzyme induced by the different ligands. From the evaluation of the data obtained, it was possible to identify combinations of molecules that gave a synergistic effect on protease Mpro. In fact, the drug combination favipiravir (6-fluoro-3,4-dihydro-3-oxo-2-pyrazinecarboxamide)/carvedilol (**2**) was found to be active against this enzyme. The inhibitory activity tests the evaluation of the synergistic effect and was evaluated by using the "3CL Protease (SARS-CoV-2) Assay Kit" from "BPS Bioscience." The best results have been obtained by mixing 2 μM compound **2** and 1 μM favipiravir showing an inhibition percentage of 98 and a contact-dependent growth inhibition (CDI) of 0.89. CF analysis (analysis of contact frequency) and MD simulation (molecular dynamics simulations) were able to identify the amino acid residues that affect the bond between compound **2** and Mpro (for example His41, Met49, and Thr25 showed more than 80% CF; Cys44, Ser46, and Glu166 more than 70%) [5]. This study supports the computational work carried out in 2020 by Zhou et al., showing the important role of carvedilol in the treatment of COVID-19 [46,78] and that of Wu et al. [17] reporting carvedilol as a potential protease inhibitor similar to 3-chymotrypsin SARS-CoV-2. Carvedilol (**2**) was also used in a clinical trial to evaluate the clinical outcomes of hypertensive patients infected with SARS-CoV-2, who commonly use inhibitors of the renin-angiotensin-aldosterone system. The study conducted by Najmeddin et al. [47] confirmed that there are no deleterious effects following the use of angiotensin-converting enzyme inhibitors (ACEis) and angiotensin receptor blockers (ARBs) in hypertensive patients with COVID-19 [48]. Also, according to the report of Onohuean et al. [9] in 2021, drugs such as carvedilol (**2**) may control the development of HF by reducing the infectivity of the 2019 novel coronavirus (SARS-CoV-2) and prevent the production of cytokine storms in severely affected COVID-19 people. Compound **2** downregulates cardiac ACE2 and inhibits SARS-CoV-2-induced acute cardiac injury [49]. Amirshahrokhi et al. demonstrated that **2** can moderate the development of paraquat-induced ALI through suppression of oxidative stress and NF-κB signaling pathway [79]. Also, **2** effectively manages Coronavirus disease 2019 complications such as esophageal varices [50] and post-COVID-19 sinus tachycardia [51]. Therefore, for its antiviral and anti-inflammatory activities, carvedilol may have dual protective effects in COVID-19 by mitigating the development of HF and ALI [52].

Several studies show the antiviral property of plant-derived molecules against RNA viruses [80]. Some of these, such as carbazole alkaloids from *Murraya koenigii* have been evaluated for SARS-CoV-2 infection. *Murraya koenigii,* known as the "Curry leaf tree" is a plant very widespread and of considerable pharmaceutical interest due to its numerous beneficial activities (antioxidant, antidiabetic, anticancer, anti-inflammatory, hepatoprotective, nephroprotective, cardioprotective, neuroprotective and antimicrobial, and antiviral activities). These activities are mainly due to the presence of compounds with a carbazole structure in its leaves, roots, and bark [81,82]. For such evidence, Wadanambi et al. in 2022 [53] evaluated the inhibitory potential of carbazole alkaloids from *Murraya koenigii* against Mpro by computational study. Using 3WL (5,6,7-trihydroxy-2-phenyl-4*H*-chromen-4-one) as a reference inhibitor, five carbazole alkaloids **3–7** (Figure 3) (koenigicine (8-methoxy-3,3,5-trimethyl-11*H*-pyrano[3,2-*a*]carbazol-9-ol) (**3**), mukonicine (9,11-dimethoxy-3,3,5-trimethyl-11*H*-pyrano[3,2-*a*]carbazole) (**4**), O-methylmurrayamine A (9-methoxy-3,3,5-trimethyl-11*H*-pyrano[3,2-*a*]carbazole) (**5**), koenine (3,3,5-trimethyl-11*H*-pyrano[3,2-*a*]carbazol-8-ol) (**6**), and girinimbine (3,11-dihydro-3,3,5-trimethyl-pyrano[3,2-*a*]carbazole) (**7**) displayed interactions in the active site of SARS-CoV-2 Mpro.

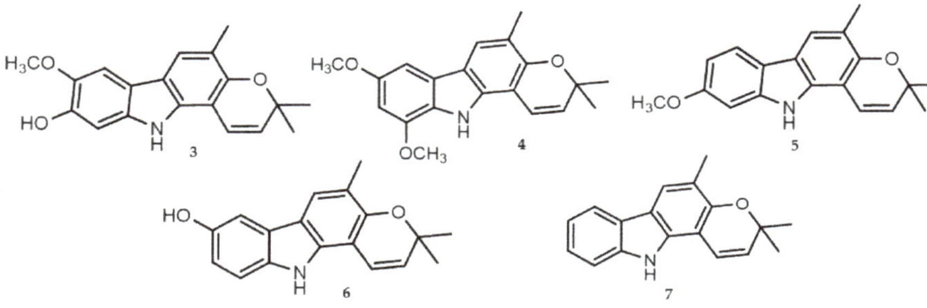

**Figure 3.** Structures of: koenigicine (**3**), mukonicine (**4**), O-methylmurrayamine A (**5**), koenine (**6**) and girinimbine (**7**).

Mainly, **4–7** may have the features to reduce SARS-CoV-2 replication by inactivating the Mpro catalytic activity. The carbazoles studied (**3–7**), compared with 3WL and showed higher binding affinity and lower binding energies towards the active site of the SC2-Mpro [83]. These compounds form hydrogen bonds with numerous aminoacid residues of the active site (for example with His41, Cys145, Asn142, etc.). In particular, the oxygen atom of the pyran ring forms a hydrogen interaction with Gly143 (for **3** and **4**), Asn142 (for **5** and **7**), and Glu166 (for **6**). Compounds **3–7** were also found to be effective against the Alpha, Beta, Gamma, and Omicron variants. Toxicity test data shows that **3**, **4**, and **5** may have carcinogenic and mutagenic effects [84], instead, **6** and **7** did not show any toxic effects to hepatotoxicity, carcinogenicity, mutagenicity, and cytotoxicity. Therefore, bioactive natural compounds **4–7**, with good oral bioavailability, represent a starting point for the synthesis of new potential SC-2 Mpro inhibitors [53].

*4.2. Viral-Entry Inhibitors Targeting Human ACE2*

Terali et al. in 2020 [54] have identified the carbazole edotecarin (**8**, Figure 4), (6-((1,3-dihydroxypropan-2-yl)amino)-2,10-dihydroxy-12-((2*R*,3*R*,4*S*,5*S*,6*R*)-3,4,5-trihydroxy-6-(hydroxymethyl)tetrahydro-2*H*-pyran-2-yl)-12,13-dihydro-5*H*-indolo[2,3-*a*]pyrrolo[3,4-*c*]carbazole-5,7(6*H*)-dione) as viral-entry inhibitors targeting human ACE2 by molecular docking.

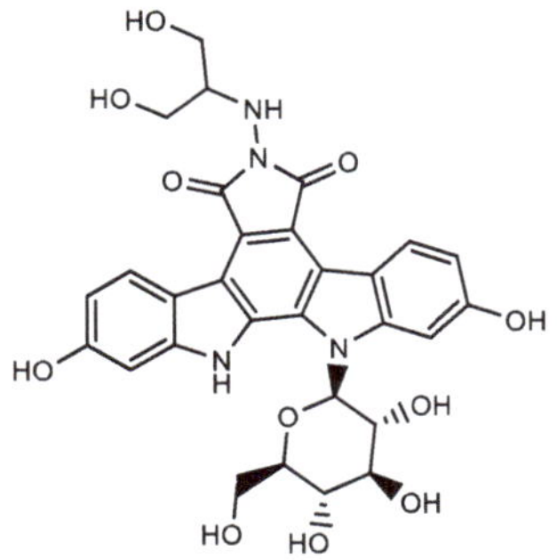

**Figure 4.** Structure of edotecarin (**8**).

Compound **8**, by means of electrostatic interactions, favors the closed (substrate/inhibitor-bound) conformation of angiotensin-converting enzyme 2 modifying the positions of the receptor's amino acids interested in recognition by SARS-CoV-2. Specifically, the diol group of topoisomerase I inhibitor (**8**) is crucial for the interaction; the catalytic residues of ACE2 mainly interested in interacting with **8** are Glu375, Tyr515, and Asn149. Furthermore, **8** with conjugative *p*-planes performs pep interactions (sandwich or T-shaped) with the amino acids Phe274, His345, and Tyr510 [50]. As already discussed, modulators of expression levels of proteins such as ACE2 may control the SARS-CoV-2 infection [55].

In 2022, Serra et al. [56] studied the synergistic effect of an analog of edotecarin (**9**, Figure 5) and bafetinib. In particular, with computational methods, they found that 7-hydroxystaurosporine (**9**), ((5*S*,6*R*,7*R*,9*R*,16*R*)-16-hydroxy-6-methoxy-5-methyl-7-(methylamino)-6,7,8,9,15,16-hexahydro-17-oxa-4*b*,9*a*,15-triaza-5,9-methanodibenzo[*b,h*] cyclonona[*jkl*]cyclopenta[*e*]-as-indacen-14(5*H*)one), and bafetinib (4-[[(3*S*)-3-(dimethyl amino)pyrrolidin-1-yl]methyl]-*N*-[4-methyl-3-[(4-pyrimidin-5-ylpyrimidin-2-yl)amino] phenyl]-3-(trifluoromethyl)benzamide), inhibit viral infection when combined together.

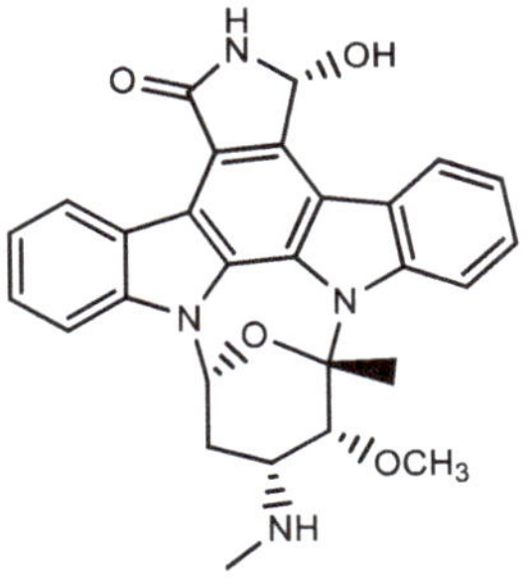

**Figure 5.** Structure of 7-hydroxystaurosporine (**9**).

In vitro studies confirmed that these compounds, used in combination, hinder a post-entry mechanism of the virus and efficacy against the Delta variant. HEK-293 T cells stably expressing human ACE2 and TMPRSS2 (HEK-293 TAT), infected with the SARS-CoV-2 strain isolated from Wuhan, were used for the experiments. The drugs were tested at concentrations of 0.09, 0.9, and 9 μM. Antineoplastic agent **9** and bafetinib (second-generation tyrosine kinase inhibitor) showed significant inhibition at 9 μM. The combination of bafetinib and 7-hydroxystaurosporine (**9**) on Caco2-ACE2 cells (Caco2 is an immortalized cell linehuman colorectal adenocarcinoma cells) has also been studied using 1 or 3 μM concentrations of drug **9** in combination with 3 μM bafetinib. At 3 μM, the treatment decreased infection by >70%. The synergistic effect was also observed for the Delta variant in a concentration-dependent manner [56,85,86]. Another staurosporine analog (**10**, Figure 6) (sodium 3-(4-((((*S*)-5-((5*S*,7*S*,8*R*,9*S*)-8-methoxy-9-methyl-16-oxo-6,7,8,9,15,16-hexahydro-5*H*,14*H*-4*b*,9*a*,15-triaza-5,9-methanodibenzo[b,h]cyclonona[jkl]cyclopenta[e]-as-indacen-7-yl)-4-oxohexanamido)methyl)-1*H*-1,2,3-triazol-1-yl)propane-1-sulfonate), called CIMSSNa, was studied in 2022 by Cheshenko et al. as SARS-CoV-2 inhibitor. Experiments conducted on three cell lines (Vero, Huh7 and Calu-3 cells) confirmed that **10** inhibits SARS-CoV-2 and that SARS-CoV-2 enters the cells by direct fusion (at least partly) [57].

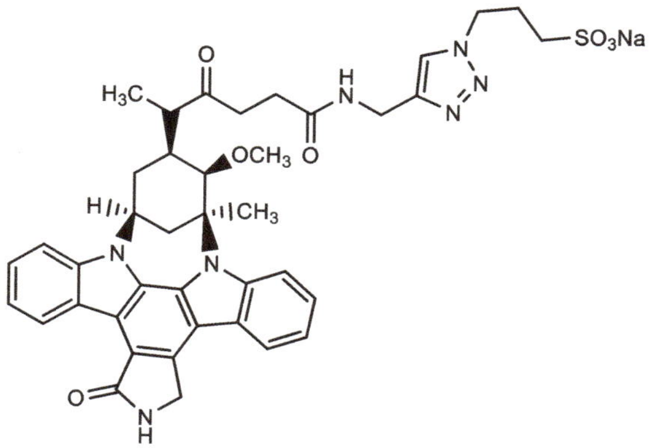

**Figure 6.** Structure of CIMSSNa (**10**).

Intead, Tanimoto et al. in 2021 [58] proved that treatment with AHR agonists (aryl hydrocarbon receptor agonists), as 6-formylindolo(3,2-*b*)carbazole (**11**, Figure 7), decreases expression of ACE2 via AHR activation, resulting in the suppression of SARS-CoV-2 disease in mammalian cells. The studies were conducted on HepG2 cells. RNA-seq analysis demonstrated that **11** increased CYP1A1 gene expression in a dose-dependent manner and inhibited the expression of the ACE2 gene. Also, the ACE2 expression in Vero E6 cells (Vero C1008 African green monkey kidney cell LIne, Clone E6) was valued [87]; again ACE2 expression is downregulated by treatment with **11**. Therefore, these results demonstrated that formylindolo carbazole **11**, the agonist of AHR, blocks the expression of ACE2 in mammalian cells, limits the entry of SARS-CoV-2, and stimulates the immune system [58].

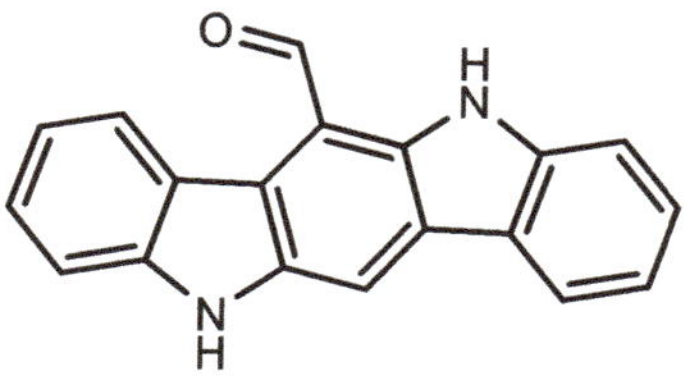

**Figure 7.** Structure of 6-formylindolo(3,2-*b*)carbazole (**11**).

### 4.3. NPC1 Inhibitor

García-Dorival et al. in 2021 [59] reported a link between the SARS-CoV-2 nucleoprotein (N) and NPC1. They have pointed out that several molecules interact with NPC1, as 2-((2-(1-benzylpiperidin-4-yl)ethyl)amino)-*N*-(9*H*-carbazol-9-yl)acetamide (**12**, Figure 8) were able to decrease SARS-CoV-2 infection with excellent selectivity in human cell infection models. In fact, **12** inhibited more than 95% of the infection of SARS-CoV-2 in Vero-E6 (Vero C1008 African green monkey kidney Cell Line, Clone E6) and A549 (epithelial cells) cells. These data suggest the importance of NPC1 for SARS-CoV-2 viral infection; NPC1, therefore, represents a potential therapeutic target to fight against SARS-CoV-2 infection [59].

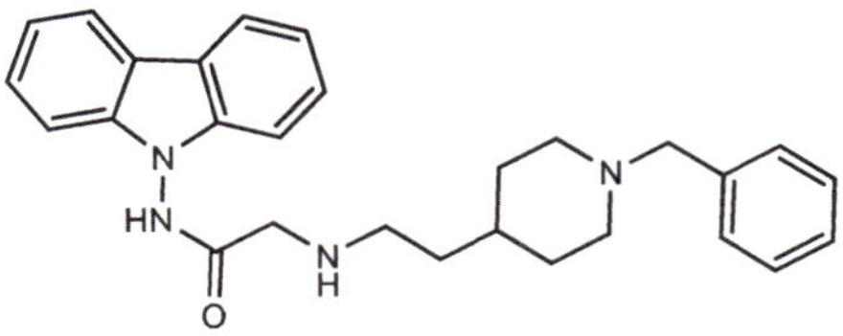

**Figure 8.** Structure of 2-((2-(1-benzylpiperidin-4-yl)ethyl)amino)-*N*-(9*H*-carbazol-9-yl)acetamide (**12**).

### 4.4. Antiviral against PLpro

Elkaeed et al. in 2022 [60] carried out computational methods such as similarity assessment, fingerprints check, docking, absorption, distribution, metabolism, excretion, toxicity (ADMET), and density-functional theory (DFT) on different metabolites of natural origin as carbazole **13** (Figure 9), (6-cyano-5-methoxy-12-methylindolo [2,3A] carbazole). This was reported as antiviral against PLpro. The binding ability against PLpro was screened through docking studies. In order to confirm the inhibitory effect of the compounds they examined against PLpro and SARS-CoV-2, other studies such as in-vitro and in-vivo studies are needed [60].

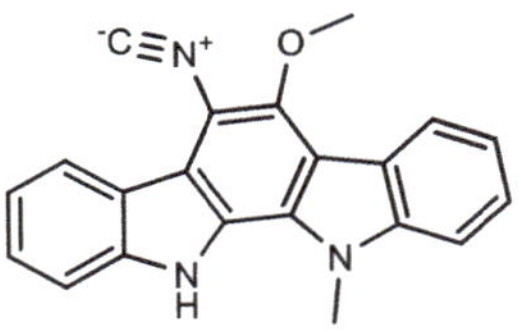

**Figure 9.** Structure of 6-cyano-5-methoxy-12-methylindolo [2,3A] carbazole (**13**).

### 4.5. Immunotherapy Treatment

As suggested by Gupta and Chiang in 2020 [70], in the development of the COVID-19 infection, an immunotherapy treatment, for example with ramatroban (**14**, Figure 10) ((3-[(3*R*)-3-[(4-fluorophenyl)sulfonylamino]-1,2,3,4-tetrahydrocarbazol-9-yl]propanoic acid)), could be necessary in case of lymphopenia, a predictor of disease severity and outcomes.

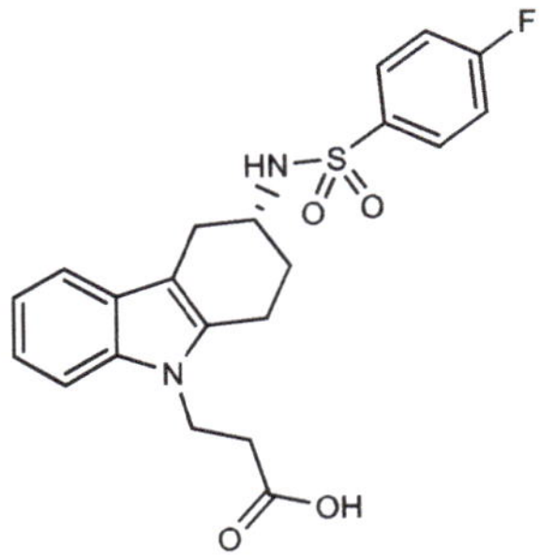

**Figure 10.** Structure of ramatroban (**14**).

Ramatroban (**14**), a selective PGD2 inhibitor and IL-13 secretion stimulator ($IC_{50}$ = 118 nM) may be needed to restore immune dysfunction during the symptomatic phase of COVID-19. Also, in 2002, Chiang et al. [61] reported in a review that **14** produces beneficial effects at all stages of SARS-CoV-2 infection as it is an immunomodulator, antithrombotic, anti-inflammatory, and antifibrotic agent. For these reasons, drug **14** gave relief of dyspnea and hypoxemia in patients with COVID-19 and, as reported in the study by Ogletree et al. [62] in 2022, it was possible to avoid hospitalization.

## 5. Conclusions

COVID-19 is a multi-organ disease involving immune deficiency and severe lung inflammation related to an inappropriate immune response and increased cytokine production. Therefore, the therapeutic methods currently being studied include the treatment with a combination of dual agents such as anti-proinflammatory cytokines and antivirals, passive immunotherapy and monoclonal antibodies, or anti-infectious and life-support therapies.

Several treatment approaches for COVID-19 have been employed since the beginning of the pandemic, such as antiviral, antimicrobial, anti-inflammatory, immunomodulator agents, antagonists of bradykinin B2 receptor, blockers of angiotensin II receptor, and corticosteroids, but many of them were found to be ineffective. Thus, the identification of effective drugs to treat COVID-19 is badly needed. Studies and observations reported in this review about treatment with carbazoles (Table 1) may represent a potentially useful clinical application, however, further investigations are required to clarify the details.

**Author Contributions:** Conceptualization, A.C.; methodology, M.G.B.; software, G.S.; resources, M.S.S. and C.S.; data curation, A.C. and M.G.B.; writing-original draft preparation, A.C. and M.G.B.; writing-review and editing, A.C. and H.E.-K.; supervision, G.S., M.S.S., and C.S. All authors have read and agreed to the published version of the manuscript.

**Funding:** This research received no external funding.

**Institutional Review Board Statement:** Not applicable.

**Informed Consent Statement:** Not applicable.

**Data Availability Statement:** Not applicable.

**Conflicts of Interest:** The authors declare no conflict of interest.

## References

1. Chen, J. Pathogenicity and Transmissibility of 2019-NcoV-A Quick Overview and Comparison with Other Emerging Viruses. *Microbes Infect.* **2020**, *22*, 69–71. [CrossRef] [PubMed]
2. Ahmad, I.; Muhammad, I.; Muhammad Waseem, M.; Basra, R.; Asim, M. Molecular Docking and Computational Exploration of Isolated Drugs from Daphne Species Against COVID-19. *Iran. J. Chem. Chem. Eng.* **2021**, *40*, 2019–2027.
3. Iacopetta, D.; Ceramella, J.; Catalano, A.; Saturnino, C.; Pellegrino, M.; Mariconda, A.; Longo, P.; Sinicropi, M.S.; Aquaro, S. COVID-19 at a glance: An up-to-date overview on variants, drug design and therapies. *Viruses* **2022**, *14*, 573. [CrossRef] [PubMed]
4. Dhama, K.; Khan, S.; Tiwari, R.; Sircar, S.; Bhat, S.; Malik, Y.S.; Singh, K.P.; Chaicumpa, W.; Bonilla-Aldana, D.K.; Rodriguez Morales, A.J. Coronavirus Disease 2019–COVID-19. *Clin. Microbiol. Rev.* **2020**, *33*, e00028-20. [CrossRef] [PubMed]
5. Abdel-Halim, H.; Hajar, M.; Hasouneh, L.; Abdelmalek, S.M.A. Identification of Drug Combination Therapies for SARS-CoV-2: A Molecular Dynamics Simulations Approach. *Drug Des. Dev. Ther.* **2022**, *16*, 2995–3013. [CrossRef] [PubMed]
6. WHO Virtual Press Conference on COVID-19—11 March 2020. Available online: https://www.who.int/docs/default-source/coronaviruse/transcripts/who-audio-emergencies-coronavirus-press-conference-full-and-final-11mar2020.pdf?sfvrsn=cb432bb3_2 (accessed on 8 September 2022).
7. Catalano, A.; Iacopetta, D.; Ceramella, J.; Maio, A.C.D.; Basile, G.; Giuzio, F.; Bonomo, M.G.; Aquaro, S.; Walsh, T.J.; Sinicropi, M.S.; et al. Are Nutraceuticals Effective in COVID-19 and Post-COVID Prevention and Treatment? *Foods* **2022**, *11*, 2884. [CrossRef] [PubMed]
8. Al-kuraishy, H.M.; Al-Gareeb, A.I.; Alblihed, M. COVID-19 and risk of acute ischemic stroke and acute lung injury in patients with type II diabetes mellitus: The anti-infammatory role of metformin. *Front. Med.* **2021**, *8*, 644295. [CrossRef]
9. Onohuean, H.; Al Kuraishy, H.M.; Al Gareeb, A.I.; Qusti, S.; Alshammari, E.M.; El Saber Batiha, G. Covid 19 and development of heart failure: Mystery and truth. *Naunyn Schmiedebergs Arch. Pharmacol.* **2021**, *394*, 2013–2021. [CrossRef]
10. Velavan, T.P.; Meyer, C.G. The COVID-19 epidemic. *Trop. Med. Int. Health* **2020**, *25*, 278–280. [CrossRef]
11. Salamanna, F.; Veronesi, F.; Martini, L.; Landini, M.P.; Fini, M. Post-COVID-19 syndrome: The persistent symptoms at the post-viral stage of the disease. A systematic review of the current data. *Front. Med.* **2021**, *8*, 653516. [CrossRef]
12. Maltezou, H.; Pavli, A.; Tsakris, A. Post-COVID syndrome: An insight on its pathogenesis. Post-COVID Syndrome: An Insight on Its Pathogenesis. *Vaccines* **2021**, *9*, 497. [CrossRef] [PubMed]
13. Fernandez-de-Las-Penas, C.; Palacios-Cena, D.; Gomez-Mayordomo, V.; Cuadrado, M.L.; Florencio, L.L. Defining post-COVID symptoms (post-acute COVID, long COVID, persistent post-COVID): An integrative classification. *Int. J. Environ. Res. Public Health* **2021**, *18*, 2621. [CrossRef]

14. Yang, H.; Yang, M.; Ding, Y. The crystal structures of severe acute respiratory syndrome virus main protease and its complex with an inhibitor. *Proc. Natl. Acad. Sci USA* **2003**, *100*, 13190–13195. [CrossRef] [PubMed]

15. Ziebuhr, J. The coronavirus replicase. *Curr. Top. Microbiol. Immunol.* **2005**, *287*, 57–94.

16. Fehr, A.R.; Perlman, S. Coronaviruses: An overview of their replication and pathogenesis. In *Coronaviruses: Methods and Protocols*; Maier, H.J., Bickerton, E., Britton, P., Eds.; Springer: New York, NY, USA, 2015; pp. 1–23.

17. Wu, C.; Liu, Y.; Yang, Y.; Zhang, P.; Zhong, W.; Wang, Y.; Wang, Q.; Xu, Y.; Li, M.; Li, X. Analysis of Therapeutic Targets for Sars-Cov-2 and Discovery of Potential Drugs by Computational Methods. *Acta Pharm. Sin. B* **2020**, *10*, 766–788. [CrossRef] [PubMed]

18. Alattar, R.; Ibrahim, T.B.H.; Shaar, S.H. Abdalla, S.; Shukri, K.; Daghfal, J.N.; Khatib, M.Y.; Aboukamar, M.; Abukhattab, M.; Alsoub, H.A.; Almaslamani, M.A.; Omrani, A.S. Tocilizumab for the treatment of severe coronavirus disease 2019. *J. Med. Virol.* **2020**, *92*, 2042–2049. [CrossRef]

19. Deng, L.; Li, C.; Zeng, Q. Arbidol combined with LPV/r versus LPV/r alone against Corona Virus Disease 2019: A retrospective cohort study. *J. Infect.* **2020**, e1–e5. [CrossRef]

20. Singh, A.K.; Singh, A.; Shaikh, A.; Singh, R.; Misra, A. Chloroquine and hydroxychloroquine in the treatment of COVID-19 with or without diabetes: A systematic search and a narrative review with a special reference to India and other developing countries. *Diabetes Metab. Syndr.* **2020**, *14*, 241–246. [CrossRef]

21. Gavriatopoulou, M.; Ntanasis Stathopoulos, I.; Korompoki, E.; Fotiou, D.; Migkou, M.; Tzanninis, I.G.; Psaltopoulou, T.; Kastritis, E.; Terpos, E.; Dimopoulos, M.A. Emerging treatment strategies for COVID 19 infection. *Clin. Experim. Med.* **2021**, *21*, 167–179. [CrossRef]

22. Dolgin. E. The race for antiviral drugs to beat COVID—And the next pandemic. *Nature* **2021**, *592*, 340–343. [CrossRef]

23. Yavuz Şimşek, S.; Komşuoğlu Çelikyurt, İ. An update of anti-viral treatment of COVID-19. *Turk. J. Med. Sci.* **2021**, *51*, 3372–3390. [CrossRef] [PubMed]

24. Aronskyy, I.; Masoudi-Sobhanzadeh, Y.; Cappuccio, A.; Zaslavsky, E. Advances in the computational landscape for repurposed drugs against COVID-19. *Drug Discov. Today* **2021**, *26*, 2800–2815. [CrossRef] [PubMed]

25. Pushpakom, S.; Iorio, F.; Eyers, P.A. Drug repurposing: Progress, challenges and recommendations. *Nat. Rev. Drug Discov.* **2019**, *18*, 41–58. [CrossRef] [PubMed]

26. Jang, W.D.; Jeon, S.; Kim, S.; Lee, S.Y. Drugs repurposed for COVID-19 by virtual screening of 6218 drugs and cell-based assay. *Proc. Nat. Acad. Sci. USA* **2021**, *118*, e2024302118. [CrossRef]

27. Kumar, A.; Sharanya, C.; Abhithaj, J.; Sadasivan, C. Drug repurposing to identify therapeutics against COVID 19 with SARS-Cov-2 spike glycoprotein and main protease as targets: An in silico study. *ChemRxiv* **2020**. [CrossRef]

28. Lazniewski, M.; Dermawan, D.; Hidayat, S.; Muchtaridi, M.; Dawson, W.K.; Plewczynski, D. Drug repurposing for identification of potential spike inhibitors for SARS-CoV-2 using molecular docking and molecular dynamics simulations. *Methods* **2022**, *203*, 498–510. [CrossRef]

29. Cao, B.; Wang, Y.; Wen, D. A trial of lopinavir–ritonavir in adults hospitalized with severe Covid-19. *N. Engl. J. Med.* **2020**, *382*, 1787–1799. [CrossRef]

30. Tong, S.; Su, Y.; Yu, Y. Ribavirin therapy for severe COVID-19: A retrospective cohort study. *Int. J. Antimicrob. Agents* **2020**, *56*, 106114. [CrossRef]

31. Farne, H.; Kumar, K.; Ritchie, A.I.; Finney, L.J.; Johnston, S.L.; Singanayagam, A. Repurposing existing drugs for the treatment of COVID-19. *Ann. Am. Thorac. Soc.* **2020**, *17*, 1186–1194. [CrossRef] [PubMed]

32. Shen, Q.; Li, J.; Zhang, Z.; Guo, S.; Wang, Q.; An, X.; Chang, H. COVID-19: Systemic pathology and its implications for therapy. *Int. J. Biol. Sci* **2022**, *18*, 386–408. [CrossRef]

33. Beran, A.; Mhanna, M.; Srour, O.; Ayesh, H.; Stewart, J.M.; Hjouj, M.; Khokher, W.; Mhanna, A.S.; Ghazaleh, D.; Khader, Y.; et al. Clinical significance of micronutrient supplements in patients with coronavirus disease 2019: A comprehensive systematic review and meta-analysis. *Clin. Nutr. ESPEN* **2022**, *48*, 167–177. [CrossRef]

34. Anand, A.V.; Balamuralikrishnan, B.; Kaviya, M.; Bharathi, K.; Parithathvi, A.; Arun, M.; Senthilkumar, N.; Velayuthaprabhu, S.; Saradhadevi, M.; Al-Dhabi, N.A. Medicinal plants, phytochemicals, and herbs to combat viral pathogens including SARS-CoV-2. *Molecules* **2021**, *26*, 1775. [CrossRef]

35. Domi, E.; Hoxha, M.; Kolovani, E.; Tricarico, D.; Zappacosta, B. The importance of nutraceuticals in COVID-19: What's the role of resveratrol? *Molecules* **2022**, *27*, 2376. [CrossRef]

36. Balboni, E.; Zagnoli, F.; Filippini, T.; Fairweather-Tait, S.J.; Vinceti, M. Zinc and selenium supplementation in COVID-19 prevention and treatment: A systematic review of the experimental studies. *J. Trace Elem. Med. Biol.* **2022**, *71*, 126956. [CrossRef]

37. Jackson, C.B.; Farzan, M.; Chen, B.; Choe, H. Mechanisms of SARS-CoV-2 entry into cells. *Nat. Rev. Mol. Cell Biol.* **2022**, *23*, 3–20. [CrossRef]

38. Lan, J.; He, X.; Ren, Y.; Wang, Z.; Zhou, H.; Fan, S.; Zhu, C.; Liu, D.; Shao, B.; Liu, T.-Y.; et al. Structural insights into the SARS-CoV-2 Omicron RBD-ACE2 interaction. *Cell Res.* **2022**, *32*, 593–595. [CrossRef]

39. Motti, M.L.; Tafuri, D.; Donini, L.; Masucci, M.T.; De Falco, V.; Mazzeo, F. The role of nutrients in prevention, treatment and Post-Coronavirus Disease-2019 (COVID-19). *Nutrients* **2022**, *14*, 1000. [CrossRef]

40. Derosa, G.; Maffioli, P.; D'Angelo, A.; Di Pierro, F. Nutraceutical approach to preventing Coronavirus Disease 2019 and related complications. *Front. Immunol.* **2021**, *12*, 582556. [CrossRef]

41. Cámara, M.; Sánchez-Mata, M.; Fernández-Ruiz, V.; Cámara, R.; Cebadera, E.; Domínguez, L. A Review of the Role of Micronutrients and Bioactive Compounds on Immune System Supporting to Fight against the COVID-19 Disease. *Foods* **2021**, *10*, 1088. [CrossRef]

42. Gasmi, A.; Tippairote, T.; Mujawdiya, P.K.; Peana, M.; Menzel, A.; Dadar, M.; Benahmed, A.G.; Bjørklund, G. The microbiotamediated dietary and nutritional interventions for COVID-19. *Clin. Immunol.* **2021**, *226*, 108725. [CrossRef] [PubMed]

43. Shakoor, H.; Feehan, J.; Al Dhaheri, A.S.; Ali, H.I.; Platat, C.; Ismail, L.C.; Apostolopoulos, V.; Stojanovska, L. Immune-boosting role of vitamins D, C, E, zinc, selenium and omega-3 fatty acids: Could they help against COVID-19? *Maturitas* **2021**, *143*, 1–9. [CrossRef]

44. Emergency Use Authorization. U.S. Food & Drug Administration. Available online: https://www.fda.gov/emergency-preparedness-and-response/mcm-legal-regulatory-and-policy-framework/emergency-use-authorization#coviddrugs (accessed on 5 December 2022).

45. Gimeno, A.; Mestres-Truyol, J.; Ojeda-Montes, M.J.; Macip, G.; Saldivar-Espinoza, B.; Cereto-Massagué, A.; Pujadas, G.; Garcia-Vallvé, S. Prediction of Novel Inhibitors of the Main Protease (M-pro) of SARS-CoV-2 through Consensus Docking and Drug Reposition. *Int. J. Mol. Sci.* **2020**, *21*, 3793. [CrossRef]

46. Vasanthakumar, N. Beta-Adrenergic Blockers as a Potential Treatment for COVID-19 Patients. *BioEssays* **2020**, *42*, 2000094–2000102. [CrossRef]

47. Najmeddin, F.; Solhjoo, M.; Ashraf, H.; Salehi, M.; Rasooli, F.; Ghoghaei, M.; Soleimani, A.; Bahreini, M. Effects of Renin-Angiotensin-Aldosterone Inhibitors on Early Outcomes of Hypertensive COVID-19 Patients: A Randomized Triple-Blind Clinical Trial. *Am. J. Hypertens.* **2021**, *34*, 1217–1226. [CrossRef]

48. Zhang, P.; Zhu, L.; Cai, J.; Lei, F.; Qin, J.J.; Xie, J.; Liu, Y.M.; Zhao, Y.C.; Huang, X.; Lin, L.; et al. Association of inpatient use of angiotensin converting enzyme inhibitors and angiotensin II receptor blockers with mortality among patients with hypertension hospitalized with COVID-19. *Circ. Res.* **2020**, *126*, 1671–1681. [CrossRef]

49. Skayem, C.; Ayoub, N. Carvedilol and COVID-19: A potential role in reducing infectivity and infection severity of SARS-CoV-2. *Am. J. Med. Sci.* **2020**, *360*, 300. [CrossRef] [PubMed]

50. Congly, S.E.; Sadler, M.D.; Abraldes, J.G.; Tandon, P.; Lee, S.S.; Burak, K.W. Practical management of esophageal varices in the context of SARS-CoV-2 (COVID19): The Alberta protocol. *Can. Liver J.* **2020**, *3*, 300–303. [CrossRef]

51. Jadhav, K.P.; Jariwala, P.V. Ivabradine versus carvedilol in the management of palpitation with sinus tachycardia among recovered COVID-19 patients. *J. Cardiol. Cardiovasc. Med.* **2020**, *1*, 10–14.

52. Servato, M.L.; Valente, F.X.; García-Moreno, L.G.; Casas, G.; Galera, R.F.; Burcet, G.; Teixidó-Tura, G.; Calabria, H.C.; González, I.F.; Rodríguez-Palomares, J.F. Intraventricular conundrum in a SARS-CoV-2–positive patient with elevated biomarkers of myocardial injury. *JACC Case Rep.* **2021**, *3*, 566–572. [CrossRef]

53. Wadanambi, P.M.; Jayathilaka, N.; Seneviratne, K.N. A Computational Study of Carbazole Alkaloids from Murraya koenigii as Potential SARS CoV 2 Main Protease Inhibitors. *Appl. Biochem. Biotechnol.* **2022**, *195*, 573–596. [CrossRef]

54. Teralı, K.; Baddal, B.; Gülcan, H.O. Prioritizing potential ACE2 inhibitors in the COVID-19 pandemic: Insights from a molecular mechanics-assisted structure-based virtual screening experiment. *J. Mol. Graph. Model.* **2020**, *100*, 107697–107706. [CrossRef] [PubMed]

55. Scialo, F.; Daniele, A.; Amato, F.; Pastore, L.; Matera, M.G.; Cazzola, M.; Castaldo, G.; Bianco, A. ACE2: Te major cell entry receptor for SARS-CoV-2. *Lung* **2020**, *198*, 867–877. [CrossRef]

56. Serra, A.; Fratello, M.; Federico, A.; Ojha, R.; Provenzani, R.; Tasnadi, E.; Cattelani, L.; del Giudice, G.; Kinaret, P.A.S.; Saarimäki, L.A.; et al. Computationally prioritized drugs inhibit SARS-CoV-2 infection and syncytia formation. *Brief. Bioinform.* **2022**, *23*, bbab507. [CrossRef] [PubMed]

57. Cheshenko, N.; Bonanno, J.B.; Hoffmann, H.H.; Jangra, R.K.; Chandran, K.; Rice, C.M.; Almo, S.C.; Herold, B.C. Cell-impermeable staurosporine analog targets extracellular kinases to inhibit HSV and SARS-CoV-2. *Commun. Biol.* **2022**, *5*, 1096–1110. [CrossRef] [PubMed]

58. Tanimoto, K.; Hirota, K.; Fukazawa, T.; Matsuo, Y.; Nomura, T.; Tanuza, N.; Hirohashi, N.; Bono, H.; Sakaguchi, T. Inhibiting SARS CoV 2 infection in vitro by suppressing its receptor, angiotensin converting enzyme 2, via aryl hydrocarbon receptor signal. *Sci. Rep.* **2021**, *11*, 16629–16640. [CrossRef] [PubMed]

59. García-Dorival, I.; Cuesta-Geijo, M.A.; Barrado-Gil, L.; Galindo, I.; Garaigorta, U.; Urquiza, J.; del Puerto, A.; Campillo, N.E.; Martínez, A.; Gastaminza, P.; et al. Identification of Niemann-Pick C1 protein as a potential novel SARS-CoV-2 intracellular target. *Antivir. Res.* **2021**, *194*, 105167–105176. [CrossRef]

60. Elkaeed, E.B.; Metwaly, A.M.; Alesawy, M.S.; Saleh, A.M.; Alsfouk, A.A.; Eissa, I.H. Discovery of Potential SARS-CoV-2 Papain-like Protease Natural Inhibitors Employing a Multi-Phase In Silico Approach. *Life* **2022**, *12*, 1407. [CrossRef]

61. Chiang, K.C.; Rizk, J.G.; Nelson, D.J.; Krishnamurti, L.; Subbian, S.; Imig, J.D.; Khan, I.; Reddy, S.T.; Gupta, A. Ramatroban for chemoprophylaxis and treatment of COVID-19: David takes on Goliath. *Expert Opin. Ther. Targ.* **2022**, *26*, 13–28. [CrossRef]

62. Ogletree, M.L.; Kulshreshta, R.; Agarwal, A.; Agarwal, A.; Gupta, A. Treatment of COVID-19 Pneumonia and Acute Respiratory Distress with Ramatroban, a Thromboxane A2 and Prostaglandin D2 Receptor Antagonist: A Four-Patient Case Series Report. *Front Pharmacol.* **2022**, *13*, 904020–904037. [CrossRef]

63. Caruso, A.; Lancelot, J.C.; El-Kashef, H.; Sinicropi, M.S.; Legay, R.; Lesnard, A.; Rault, S. A Rapid and Versatile Synthesis of Novel Pyrimido[5,4-b]carbazoles. *Tetrahedron* **2009**, *65*, 10400–10405. [CrossRef]

64. Iacopetta, D.; Rosano, C.; Puoci, F.; Parisi, O.I.; Saturnino, C.; Caruso, A.; Longo, P.; Ceramella, J.; Malzert-Fréon, A.; Dallemagne, P.; et al. Multifaceted Properties of 1,4-Dimethylcarbazoles: Focus on Trimethoxybenzamide and Trimethoxyphenylurea Derivatives as Novel Human Topoisomerase II Inhibitors. *Eur. J. Pharm. Sci.* **2017**, *96*, 263–272. [CrossRef] [PubMed]

65. Caruso, A.; Chimento, A.; El-Kashef, H.; Lancelot, J.C.; Panno, A.; Pezzi, V.; Saturnino, C.; Sinicropi, M.S.; Sirianni, R.; Rault, S. Antiproliferative Activity of Some 1,4-Dimethylcarbazoles on Cells That Express Estrogen Receptors: Part, I. *J. Enzym. Inhib. Med. Chem.* **2012**, *27*, 609–613. [CrossRef] [PubMed]

66. Caruso, A.; Voisin-Chiret, A.S.; Lancelot, J.C.; Sinicropi, M.S.; Garofalo, A.; Rault, S. Efficient and Simple Synthesis of 6-Aryl-1,4-Dimethyl-9H-Carbazoles. *Molecules* **2008**, *13*, 1312–1320. [CrossRef] [PubMed]

67. Sinicropi, M.S.; Iacopetta, D.; Rosano, C.; Randino, R.; Caruso, A.; Saturnino, C.; Muià., N.; Ceramella, J.; Puoci, F.; Rodriquez, M.; et al. N-thioalkylcarbazoles derivatives as new anti-proliferative agents: Synthesis, characterisation and molecular mechanism evaluation. *J. Enzym. Inhib. Med. Chem.* **2018**, *33*, 434–444. [CrossRef] [PubMed]

68. Caruso, A.; Barbarossa, A.; Carocci, A.; Salzano, G.; Sinicropi, M.S.; Saturnino, C. Carbazole derivatives as STAT inhibitors: An overview. *Appl. Sci.* **2021**, *11*, 6192. [CrossRef]

69. Saturnino, C.; Iacopetta, D.; Sinicropi, M.S.; Rosano, C.; Caruso, A.; Caporale, A.; Marra, N.; Marengo, B.; Pronzato, M.A.; Parisi, O.I.; et al. N-alkyl carbazole derivatives as new tools for Alzheimer's disease: Preliminary studies. *Molecules* **2014**, *19*, 9307–9317. [CrossRef]

70. Gupta, A.; Chiang, K.C. Prostaglandin D2 as a mediator of lymphopenia and a therapeutic target in COVID-19 disease. *Med. Hypotheses* **2020**, *143*, 110122–110124. [CrossRef]

71. Gimeno, A.; Ojeda-Montes, M.; Tomás-Hernández, S.; Cereto-Massagué, A.; Beltrán-Debón, R.; Mulero, M.; Pujadas, G.; Garcia-Vallvé, S. The Light and Dark Sides of Virtual Screening: What Is There to Know? *Int. J. Mol. Sci.* **2019**, *20*, 1375. [CrossRef]

72. Mody, V.; Ho, J.; Wills, S.; Mawri, A.; Lawson, L.; Ebert, M.C.; Fortin, G.M.; Rayalam, S.; Taval, S. Identifcation of 3-chymotrypsin like protease (3CLPro) inhibitors as potential anti-SARSCoV-2 agents. *Commun. Biol.* **2021**, *4*, 93. [CrossRef]

73. Shin, D.; Mukherjee, R.; Grewe, D.; Bojkova, D.; Baek, K.; Bhattacharya, A.; Schulz, L.; Widera, M.; Mehdipour, A.R.; Tascher, G.; et al. Papain-like protease regulates SARS-CoV-2 viral spread and innate immunity. *Nature* **2020**, *587*, 657–662. [CrossRef]

74. Báez-Santos, Y.M.; John, S.E.S.; Mesecar, A.D. The SARS-coronavirus papain-like protease: Structure, function and inhibition by designed antiviral compounds. *Antivir. Res.* **2015**, *115*, 21–38. [CrossRef] [PubMed]

75. Zhang, L.; Lin, D.; Sun, X.; Curth, U.; Drosten, C.; Sauerhering, L.; Becker, S.; Rox, K.; Hilgenfeld, R. Crystal structure of SARS-CoV-2 main protease provides a basis for design of improved $\alpha$-ketoamide inhibitors. *Science* **2020**, *368*, 409–412. [CrossRef] [PubMed]

76. Kong, R.; Yang, G.; Xue, R.; Liu, M.; Wang, F.; Hu, J.; Guo, X.; Chang, S. COVID-19 Docking Server: A meta server for docking small molecules, peptides and antibodies against potential targets of COVID-19. *Bioinformatics* **2020**, *36*, 5109–5111. [CrossRef] [PubMed]

77. Tang, B.; He, F.; Liu, D.; He, F.; Wu, T.; Fang, M.; Niu, Z.; Wu, Z.; Xu, D. AI-aided design of novel targeted covalent inhibitors against SARS-CoV-2. *Biomolecules* **2022**, *12*, 746. [CrossRef]

78. Zhou, Y.; Hou, Y.; Shen, J.; Huang, Y.; Martin, W.; Cheng, F. Network-based drug repurposing for novel coronavirus 2019-nCoV/SARS-CoV-2. *Cell Discov.* **2020**, *6*, 14. [CrossRef]

79. Amirshahrokhi, K.; Khalili, A.R. Carvedilol attenuates paraquatinduced lung injury by inhibition of proinfammatory cytokines, chemokine MCP-1, NF-κB activation and oxidative stress mediators. *Cytokine* **2016**, *88*, 144–153. [CrossRef]

80. Owen, L.; Laird, K.; Shivkumar, M. Antiviral plant-derived natural products to combat RNA viruses: Targets throughout the viral life cycle. *Lett. Appl. Microbiol.* **2021**, *75*, 476–499. [CrossRef]

81. Balakrishnan, R.; Vijayraja, D.; Jo, S.H.; Ganesan, P.; Su-Kim, I.; Choi, D.K. Medicinal profle, phytochemistry, and pharmacological activities of Murraya koenigii and its primary bioactive compounds. *Antioxidants* **2020**, *9*, 101. [CrossRef]

82. Caruso, A.; Ceramella, J.; Iacopetta, D.; Saturnino, C.; Mauro, M.V.; Bruno, R.; Aquaro, S.; Sinicropi, M.S. Carbazole Derivatives as Antiviral Agents: An Overview. *Molecules* **2019**, *24*, 1912. [CrossRef]

83. Schultes, S.; De Graaf, C.; Haaksma, E.E.; De Esch, I.J.; Leurs, R.; Krämer, O. Ligand efciency as a guide in fragment hit selection and optimization. *Drug Discov. Today Technol.* **2010**, *7*, e157–e162. [CrossRef]

84. Guo, Z. The modifcation of natural products for medical use. *Acta Pharm. Sin. B.* **2017**, *7*, 119–136. [CrossRef] [PubMed]

85. Sampath, D.; Cortes, J.; Estrov, Z.; Du, M.; Shi, Z.; Andreeff, M.; Gandhi, V.; Plunkett, W. Pharmacodynamics of cytarabine alone and in combination with 7- hydroxystaurosporine (UCN-01) in AML blasts in vitro and during a clinical trial. *Blood* **2006**, *107*, 2517–2524. [CrossRef]

86. Santos, F.P.S.; Kantarjian, H.; Cortes, J.; Quintas-Cardama, A. Bafetinib, a dual BcrAbl/Lyn tyrosine kinase inhibitor for the potential treatment of leukemia. *Curr. Opin. Investig. Drugs* **2010**, *11*, 1450–1465. [PubMed]

87. Matsuyama, S.; Nao, N.; Shirato, K.; Kawase, M.; Saito, S.; Takayama, I.; Nagata, N.; Sekizuka, T.; Katoh, H.; Kato, F.; et al. Enhanced isolation of SARS-CoV-2 by TMPRSS2-expressing cells. *Proc. Natl. Acad. Sci. USA* **2020**, *117*, 7001–7003. [CrossRef] [PubMed]

MDPI

St. Alban-Anlage 66

4052 Basel

Switzerland

Tel. +41 61 683 77 34

Fax +41 61 302 89 18

www.mdpi.com

*Applied Sciences* Editorial Office

E-mail: applsci@mdpi.com

www.mdpi.com/journal/applsci